More Than One Mystery

Tom Oses

Mark P. Silverman

More Than One Mystery
Explorations in Quantum Interference

With 51 Illustrations

Springer-Verlag
New York Berlin Heidelberg London Paris
Tokyo Hong Kong Barcelona Budapest

Mark P. Silverman
Department of Physics
Trinity College
Hartford, CT 06106
USA

Library of Congress Cataloging-in-Publication Data
Silverman, Mark P.
    More than one mystery : explorations in quantum interference /
Mark P. Silverman.
        p.   cm.
    Includes bibliographical references and index.
    ISBN 0-387-94340-4. — ISBN 3-540-94340-4 (hardcover)
    ISBN 0-387-94376-5. — ISBN 3-540-94376-5 (softcover)
    1. Quantum theory. I. Title.
QC174.12.S525 1994
530.1′2—dc20   94-21442

Printed on acid-free paper.

Production coordinated by Brian Howe and managed by Bill Imbornoni; manufacturing
supervised by Gail Simon.
Typeset by Integral Typesetting, UK.
Printed and bound by Hamilton Printing Co., Castleton, NY.
Printed in the United States of America.

9 8 7 6 5 4 3 2 1

ISBN 0-387-94340-4 Springer-Verlag New York Berlin Heidelberg (hardcover)
ISBN 3-540-94340-4 Springer-Verlag Berlin Heidelberg New York
ISBN 0-387-94376-5 Springer-Verlag New York Berlin Heidelberg (softcover)
ISBN 3-540-94376-5 Springer-Verlag Berlin Heidelberg New York

To Susan, Chris, and Jennifer
and to the memory of
Richard P. Feynman

Preface

There was a time when the newspapers said that only twelve men understood the theory of relativity. I do not believe there ever was such a time. There might have been a time when only one man did, because he was the only guy who caught on, before he wrote his paper. But after people read the paper, a lot of people understood the theory of relativity in some way or other, certainly more than twelve. On the other hand, I think I can safely say that nobody understands quantum mechanics.

The Character of Physical Law
Richard Feynman

A few years ago I had the distinction and pleasure of presenting the Erskine Lectures at the University of Canterbury in Christchurch, New Zealand, as lovely a country and as hospitable a university as one might hope to find anywhere. Asked at the time of invitation—yet before official sanction by the university administration—to provide an overall title for the talks "using only words that a Vice-Chancellor may be expected to understand or have heard of," I selected the title proposed by my host of "Novel Quantum Interference Effects in Particle, Atomic, and Chemical Physics." Although the lectures actually embraced topics more diverse than quantum mechanics, I noted in my reply that, for the purpose of convincing a Vice-Chancellor, the title was probably accurate enough since the explanation of any phenomenon, at a level sufficiently deep, undoubtedly involved quantum mechanics.

Presumptuous as it may be, in view of Feynman's remarks cited above—and who can pretend to have understood quantum mechanics better than he?—to lecture on a topic that no one understands, I was consoled and encouraged by the thought that at least I was in good company, for even the great Michael Faraday had written of himself:

As when on some secluded branch in forest far and wide sits perched an owl, who, full of self-conceit and self-created wisdom, explains, comments, condemns, ordains and orders things not understood, yet full of his importance still holds forth to stocks and stones around—so sits and scribbles Mike.[1]

That describes, I suppose, the occupational condition of professors in general, whatever their field of interest. Nevertheless, with respect to the subject at hand, I do not think it unreasonable to say that, Feynman's comment notwithstanding, quantum mechanics is a tool for most physicists—and one can use a tool to great effect without knowing the mysteries of how it was produced or why it may work exactly as it does. Thus can quantum mechanics provide deep understanding of the physical world without being totally comprehensible itself.

Also, just as one can understand Einstein's theory of relativity "in some way or other" as Feynman says, so, too, can one try to grasp quantum mechanics in some way or other. Most physicists of course have an image in their minds, however incomplete or inadequate, of the mechanisms underlying the processes they are studying. The world is too complex to fathom in its entirety—and much the same can be said for even a small part of it. The art of science, I believe, is constructing models. And the *art* of this art is to have at hand the simplest models which provide useful, thought-provoking explanations unencumbered by extraneous detail. (The more one delves, of course, the less extraneous some of the details may become, but that is how science progresses.)

The essence of my Erskine Lectures, which were directed to a mix of faculty, research staff, and students (both graduate and undergraduate), was in part to demonstrate the effectiveness of a variety of simple model systems in treating problems that interested me, and to which I gave much thought, over a period of some twenty-five years or more. In choosing problems to work on, whether of quantum mechanical origin or not, I have always tried to steer a sort of middle course between pure utilitarianism at one end and philosophical speculation devoid of observable consequence at the other. Thus, the lectures addressed conceptually interesting questions concerning, among other things, puzzling phenomena arising from quantum interference, the indistinguishability of identical particles, the "ghostly" (in Einstein's words) nonlocal interactions of particles and fields, as well as practical applications or implications in areas of interferometry, microscopy, and spectroscopy.

The lectures, I believe, were well received—or at least the audiences were kind enough to let me think that—and from early on, and throughout the duration of my visit which lasted several months, there were requests to have a finished set of lecture notes. I intended to honor this request, but as is often the case, urgencies of the moment take precedence over reminiscences of the past, and the project proceeded in so halting a manner that even well after the time of my departure little progress had been made.

---

[1] R.L. Weber, *A Random Walk in Science* (Crane, Russak, New York, 1973), p. 76.

One principal cause of delay, in fact, was the fulfillment of an earlier obligation to complete a personal, nontechnical (or, at least, not too technical) account of my broader scientific interests—in the hope, as I eventually wrote, of demonstrating "that the study of physics can be intensely interesting and satisfying even when one is not addressing such ultimate questions as the origin and evolution of the universe." This book—*And Yet It Moves: Strange Systems and Subtle Questions in Physics*—has since been published[2] and apparently has sparked much interest. Among many encouraging comments from familiar colleagues, as well as from readers unknown to me, was an oft-repeated request to have a more mathematical treatment of the contents, particularly the portions of the book devoted to quantum mechanics. Sir Brian Pippard of the Cavendish Laboratory expressed the sentiment aptly in his generous book review when he wrote:[3] ". . . at this level of physics mathematics is a positive aid, not an encumbrance."

The present book, therefore, was born of the desire to fulfill two obligations seen largely as one, namely to make available in one volume a satisfyingly thorough—but not overwhelming—treatment of a small number of fundamental quantum systems and processes drawn from my unpublished lectures or papers scattered throughout numerous journals and conference proceedings. This is *not*, by any means, ancient material devoid of current interest. Indeed, the model systems discussed in this book are at the core of much of contemporary physics, and no doubt will remain so in the physics of the future—at least as long as quantum mechanics itself remains of interest.

Who can count, for example, the multifarious physical applications for which the "two-level atom" is a suitable model? Or when would physics ever dispense with the model of the "two-slit interferometer"? Through the vehicle of simple model systems I discuss a wide range of basic issues relating to dynamics and symmetry, interference and spectroscopy, the exclusion principle and electron microscopy, the "ghostly" correlations of entangled quantum states, the nonlocal influences of electromagnetic fields, or the physical implications of molecular chirality. Some of the systems or processes in this book have been investigated experimentally, but for others the first experimental steps are only just now being taken. Among these are to observe by electron interferometry the fermionic counterpart to correlations that Hanbury Brown and Twiss observed with light, or to test gauge invariance by measuring the Aharonov–Bohm phase shift with structured particles like ions rather than with "point" electrons, or to detect the influence of the diurnal rotation of the Earth on optical properties of atoms. In quantum mechanics, old models, if they are good ones, never die—and, unlike soldiers, never even fade away, but continually find new applications.

---

[2] M.P. Silverman, *And Yet It Moves: Strange Systems and Subtle Questions in Physics* (Cambridge University Press, London, 1993).

[3] B. Pippard, Case of Curiosities, *Nature*, **364**, 769 (1993).

Although this book discusses key aspects of quantum mechanics—bound state structure, resonant and virtual transitions, forward scattering, diffraction, and perturbation theory, among others—the principal unifying theme that runs through the chapters is that of matter wave interference. Quantum interference, that hauntingly strange process like none other in classical physics, is almost the defining attribute of quantum physics. Richard Feynman (once again), whose off-hand sayings seem to influence in so many ways the imagery of physics, proclaimed in his famous *Lectures* that he was going to:

> examine a phenomenon which is impossible, *absolutely* impossible, to explain in any classical way, and which has in it the heart of quantum mechanics. In reality, it contains the *only* mystery.[4]

and then proceeded to discuss the phenomenon of electron self-interference— a phenomenon that assumes a pivotal role in this book, too. But central as it is, I cannot agree with his assertion. As one directs attention away from systems of single particles to systems of correlated particles or particles and space together, other mysteries equally profound arise.

The counterintuitive, long-range influence of one object on another, such as first underscored by Einstein, presents an enigma of a rather different nature than the emergence of a coherent interference pattern from the random arrival of independent particles. The perplexing (from the standpoint of interpretation) physical influence of spatial topology in the nonlocal interaction of charged matter with electromagnetic fields, as emphasized by Aharonov and Bohm, has been a controversial topic for over forty years. The strange connection between spin and statistics, which occupied Pauli for much of his professional life, is yet another enigma in that the justification of this relation can only be made through an intricate and complex analysis outside the strict domain of quantum mechanics. Feynman, himself, has said of it: "You know . . . I couldn't reduce it to the freshman level. That means we really don't understand it."[5] There is *more* than one mystery in the singular and intriguing world of quantum mechanics—and all of these have a bearing on matter wave interference.[6]

Part of the enjoyment of "doing" physics comes from being able to share one's ideas and enthusiasm with others similarly inclined. In thinking about the material for this book I have benefited from exchanges with numerous colleagues around the world, and would particularly like to acknowledge the thought-provoking and agreeable discussions—not only in

---

[4] R.P. Feynman, R.B. Leighton, and M. Sands, *The Feynman Lectures on Physics*, Vol. III (Addison-Wesley, Reading, MA, 1965), p. I–1. The italic words in the citation correspond to italics in the original.

[5] D.L. Goodstein, Richard P. Feynman, Teacher, *Physics Today*, **42**, No. 2, 75 (1989).

[6] M.P. Silverman, *More Than One Mystery: Quantum Interference with Correlated Charged Particles and Magnetic Fields*, *Amer. J. Phys.*, **61**, 514 (1993).

their offices but as a guest in their homes—with Professors Jacques Badoz of the Ecole Supérieure de Physique et Chimie [ESPCI] (Paris), David Hestenes and John Spence of the Arizona State University (Tempe), and Geoffrey Stedman of the University of Canterbury (Christchurch), and Dr. Franz Hasselbach of the University of Tübingen. I am also grateful to the Hitachi Advanced Research Laboratory (Tokyo) for the hospitable and stimulating environment in which a number of my ideas on electron interferometry ripened during my invited stays as Chief Researcher in the 1980s, and to Professor Pierre de Gennes, Director of the ESPCI, for the multiple appointments to the Joliot Chair of Physics, during which tenure I explored the intricacies of chiral systems.

I would very much like to thank Dr. Thomas von Foerster, Senior Physics Editor at Springer-Verlag, for his patient encouragement, advice, and editorial skills.

As to patience and encouragement, however, I owe the greatest debt of gratitude to my wife and children, to whom this book is dedicated, for their love and understanding during intense periods of research when physically I may have been close by, but in thought far away.

Finally, it is a pleasure to pay homage to the memory of Richard Feynman whose incisive explanations have illuminated virtually all of physics. I have not always agreed with him, but I have always learned from him. Indeed, he was in a way my foremost teacher of quantum mechanics. As a Harvard student years ago, it was Julian Schwinger's elegant, if not exquisite, lectures that I attended—but it was Feynman's books and papers that I read. To him, too, this book is dedicated.

Contents

Preface vii

CHAPTER 1
Fields Without Forces 1

1.1. The Enigma of Quantum Interference 1
1.2. Confined Fields and Electron Interference 8
References 19

CHAPTER 2
Around and Around: The Rotating Electron in Electromagnetic Fields 21

2.1. Broken Symmetry of the Charged Planar Rotator 21
2.2. The Two-Dimensional Rotator in an Electric Field 24
2.3. The Two-Dimensional Rotator in a Magnetic Field 35
2.4. The Two-Dimensional Rotator in a Vector Potential Field . . . 41
2.5. Fermions, Bosons, and Things In-Between 48
2.6. Quantum Interference in a Metal Ring 52
References 56

CHAPTER 3
Interferometry of Correlated Particles 59

3.1. Ghostly Correlations: Wave and Spin 59
3.2. The AB–EPR Effect with Two Solenoids 68
3.3. The AB–HBT Effect in a Two-Slit Interferometer 75
3.4. Correlated Particles in a Mach–Zehnder Interferometer 79
3.5. Brighter Than a Million Suns: Electron Beams from Atom-Size Sources . 90
References 97

CHAPTER 4
Quantum Boosts and Quantum Beats 100

4.1. Interfering Pathways in Time 100
4.2. Laser-Generated Quantum Beats 104

4.3. Nonlinear Effects in a Three-Level Atom 109
4.4. Correlated Beats from Entangled States 121
References 127

CHAPTER 5
Sympathetic Vibrations: The Atom in Resonant Fields . . . 129

5.1. Beams, Bottles, and Electric Resonance 129
5.2. Two Perspectives of the Two-Level Atom 139
5.3. Oscillating Field Theory 147
5.4. Resonance and Coherent States: The Tell-Tale Mark of a Quantum
Jump 156
5.5. Quantum Interference in Separated Oscillating Fields 164
5.6. Ion Interferometry and Tests of Gauge Invariance 170
References 180

CHAPTER 6
The Quantum Physics of Handedness 184

6.1. Optical Activity of Mirror-Image Molecules 184
6.2. Quantum Interference and Parity Conservation 189
6.3. Optical Activity of Rotating Matter 198
References 206

Index 209

CHAPTER 1

Fields Without Forces

Let us probe the silent places, let us seek what luck betide us
Let us journey to a lonely land I know
There's a whisper on the night-wind,
There's a star agleam to guide us,
And the Wild is calling, calling ... let us go.

The Call of the Wild
Robert Service

1.1. The Enigma of Quantum Interference

In the more than seventy years that have passed since Louis de Broglie first proposed the wave-like behavior of particles, the idea of particle interference has become more-or-less familiar to most physicists. The term "wave-particle duality," which once may have rung like a wild oxymoron, for a long while now has been an integral part of physicists' working vocabulary. And yet, despite the fact that technological advances have so facilitated the experimental demonstration of the wave-like attributes of matter, these processes are no more visualizable or explicable today than when they first reached the consciousness of the physics community. The interference of massive particles remains an intriguing phenomenon which, in the oft-cited words of Richard Feynman [1], "has in it the heart of quantum mechanics." "In reality," he wrote, "it contains the only mystery."

Let us take stock briefly of just how alien the process of particle interference is to our ordinary experience. In one recent demonstration [2] by researchers at the Hitachi Advanced Research Laboratory in Tokyo, a field-emission electron microscope was employed to produce an electron version of Young's two-slit experiment with light as shown in Figure 1.1. Electrons, drawn from a sharp tungsten filament by an applied electrostatic potential of about 3–5 kV, were subsequently accelerated through a potential difference of 50 kV to a speed of approximately 0.41 the speed of light. Subsequently split by an electrostatic device known as a Möllenstedt biprism

1

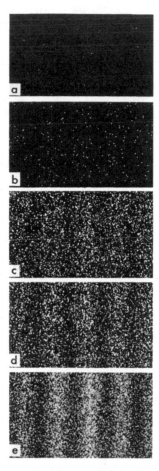

Figure 1.2. Evolution of electron interference pattern in time. Electrons arrive at the rate of approximately 1000 per second. The number recorded in each frame is: (a) 10; (b) 100; (c) 3000; (d) 20,000; and (e) 70,000. (Courtesy of A. Tonomura, Hitachi ARL.)

Indeed, except for the marked difference in wavelengths—about 500 nm for visible light and 0.005 nm (a tenth the diameter of a hydrogen atom) for the 50 kV electrons—the uninformed observer could not tell whether the fringe pattern was created by light or by particles. If the location of each electron arrival is random, and there is no communication between electrons, how then can the overall spatial distribution of detected electrons manifest a coherently organized pattern? *That* is the enigma of quantum mechanics to which Feynman referred.

Since the two-slit electron interference experiment is conceptually the simplest, if archetypal, example of quantum interference, it is worth examining it quantitatively in more detail, if only to introduce geometric and dynamical quantities that will be encountered again later. The speed β (relative to that of light) of an electron in the beam is deducible from the relativistic expression for energy conservation

$$E = mc^2/\sqrt{1 + \beta^2} = mc^2 + eV, \tag{1a}$$

and takes the form

$$\beta = \sqrt{\left(1 + \frac{eV}{mc^2}\right)^2 - 1} \Big/ \left(1 + \frac{eV}{mc^2}\right), \tag{1b}$$

where m is the electron mass, e is the magnitude of the electron charge, and V is the accelerating potential. From relation (1b), the de Broglie wavelength λ of the electron, a measure of the magnitude of the linear momentum p,

$$p = \frac{mv}{\sqrt{1 - \beta^2}} = h/\lambda, \tag{1c}$$

can be expressed in the form

$$\lambda = \frac{\lambda_C}{\sqrt{(1 + eV/mc^2)^2 - 1}}, \tag{1d}$$

where λ_C is the electron Compton wavelength

$$\lambda_C = h/mc = 2.43 \times 10^{-10} \text{ cm}. \tag{1e}$$

and h is Planck's constant.

Ideally, the electron wavelength refers to monoenergetic electrons in much the same way as an optical wavelength characterizes perfectly monochromatic light. These are idealizations that are only imperfectly realized in nature. Relying again on optical imagery (which can be misleading if taken too literally), one can describe the electron beam more appropriately in terms of wave packets in Figure 1.3. If the energy uncertainty (or

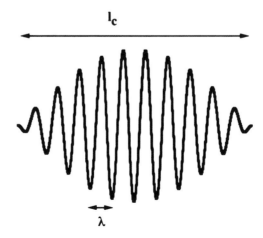

Figure 1.3. Schematic diagram of an electron wave packet. The de Broglie wavelength λ is determined by the particle linear momentum. The longitudinal coherence length l_c is determined by the dispersion in particle energy.

energy dispersion) of the beam is ΔE, then by the Heisenberg uncertainty principle there is a characteristic time interval, the beam coherence time t_c

$$t_c \sim \hbar/\Delta E, \tag{2a}$$

(with $\hbar \equiv h/2\pi$) over which an electron wave packet emerges from the source. Propagating at a mean speed v, the wave packet has a characteristic length, or longitudinal coherence length

$$l_c = vt_c. \tag{2b}$$

For a 50 kV beam with dispersion of 0.1 eV, the coherence time and length are, respectively, $t_c = 6.6 \times 10^{-15}$ s and $l_c = 7.9 \times 10^{-5}$ cm. In order for wave packets to overlap and interfere, their optical path length difference between source and detector must not be much in excess of l_c. The coherence length can greatly exceed the de Broglie wavelength—and does so in the present example by five orders of magnitude.

In addition to a longitudinal extension, the wave packets also have a lateral extension as characterized by the transverse coherence length l_t

$$l_t \sim \lambda/2\alpha, \tag{2c}$$

arising from the finite size of the source. Here 2α is the angular diameter of the source as seen from the diffracting object; equivalently, 2α is the beam divergence angle as seen from the viewing plane. To understand the relevant geometrical relations and the origin of expression (2c), consider Figure 1.4. Electrons emitted from the center of the source give rise to a diffraction

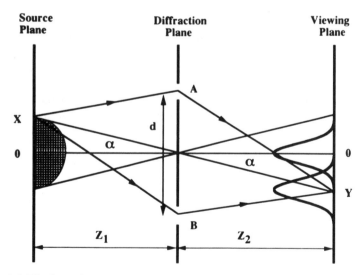

Figure 1.4. The lateral or transverse coherence length l_t is determined by the angular dispersion of the beam. It is a measure of the maximum extent of wave front separation—here equivalent to slit separation d—at which a diffracting object is coherently illuminated by an extended source.

pattern centered about the symmetry axis, i.e., at the origin O of the viewing screen. Electrons emitted from an above-axis point X in the source plane produce a diffraction pattern centered at the below-axis point Y on the viewing screen. As in light optics, points X and Y are connected by a straight line through the center of the diffracting object (in this case the midpoint between two slits a distance d apart); the corresponding distances from the symmetry axis stand in the same ratio as the separations Z_1 between the source and diffraction plane and Z_2 between the diffraction plane and viewing screen. One can readily show that the relative phase of the electron wave functions describing electrons arriving at point Y from the origin (O) and periphery (X) of the source is approximately $2\pi\, dX/\lambda Z_1$. If this phase difference is π radians, the "crests" of the waves at Y from source point X overlap the "troughs" of the waves at Y from the source point O—and the contrast or visibility of the interference fringes is zero. This defines (at least approximately) a maximum slit separation $d = \lambda/(2X/Z_1)$—to be identified with the lateral coherence width—below which the diffracting object is coherently illuminated by the extended source and gives rise to visible interference fringes. The ratio $2X/Z_1$ (for $X \ll Z_1$) is the angular diameter of the source appearing in relation (2c).

The above heuristic argument can be made more rigorous by actually integrating the diffraction pattern at the viewing screen over all contributing points of the source. In the resulting theoretical intensity distribution, the oscillatory term is multiplied by a visibility function which, in the case of a diffracting screen with two slits, has the form [3]

$$\mathscr{V} = |\sin(\pi d/l_t)|/(\pi d/l_t), \tag{2d}$$

where l_t is again the lateral coherence width of relation (2c). It is seen that $\mathscr{V}$ vanishes for $d = l_t$. In any event, the point of importance is that, as a general condition for diffraction, the transverse coherence length must not be greatly inferior to the size of the diffracting object or aperture. In the Hitachi experiment under discussion, the divergence angle was of the order of 4×10^{-8} radians, and therefore $l_t \sim 140\ \mu m$, a value larger than the approximately 0.5 μm radius of the anodic filament of the biprism, yet much smaller than the 10 mm separation of the two adjacent grounded electrodes.

Upon traversing the biprism, the components of the incident electron beam passing to one side or the other of the filament are deflected toward the observation plane and overlap there at an angle twice the deflection angle α thereby giving rise to interference fringes with fringe space $\lambda/2\alpha$. The resulting fringe spacing of around 700 nm was subsequently magnified 2000 times by a projector lens.

The details of the detection process, which involved the use of sophisticated image processing apparatus for recording the number and locations of all electrons arriving within the field of view will be left to the original literature. Suffice it to say that close to 1000 electrons arrived at the detector each second, and that a pattern of sharp fringes could be created in about one-half hour. However, at an electron speed of $\sim 0.41c$ there is a mean spatial separation of approximately 1.23×10^5 m between two sequential electrons detected 1 ms apart. At any given moment during the experiment, therefore, there is likely to be only one particle traversing the apparatus. This feature must perplex anyone seeking to understanding experiments of this kind at a deeper level than simply being able to predict the outcome (which, of course, is what quantum mechanics enables one to do).

So subtle and contrary to ordinary experience are the implications of electron self-interference, that one is usually not fully aware of the alternation in use of language required to describe the experiment. For example, to explain the action of the biprism one speaks of the deflection of the components of the beam to one side or the other of the filament. But where is there really a *beam*, for effectively only one particle at a time moves through the biprism? To be sure, knowing the spatial variation of the electrostatic field of the filament, one can calculate the deflection angle α of an incident electron. However, the supposition that an electron has with certainty actually taken one of two classically conceivable pathways through the apparatus theoretically leads to no interference effect, and, indeed, the experimental capacity to produce an interference pattern would be destroyed by the intrusive observation to test that supposition. Conversely, one can speak of the diffraction of waves around the filament. But where is there really a *wave*, for the electron is always produced and detected as an elementary corpuscular entity with discrete electric charge (4.8×10^{-10} esu), mass (9.11×10^{-28} g), and spin angular momentum ($\frac{1}{2}\hbar$).

It is frequently said or implied that the wave-particle duality of matter embodies the notion that a particle—the electron, for example—propagates like a wave, but registers at a detector like a particle. Here one must again exercise care in expression so that what is already intrinsically difficult to understand is not made more so by semantic confusion. The manifestations of wave-like behavior are statistical in nature and *always* emerge from the collective outcome of many electron events. In the present experiment nothing wave-like is discernible in the arrival of single electrons at the observation plane. It is only after the arrival of perhaps tens of thousands of electrons that a pattern *interpretable* as wave-like interference emerges. Likewise, there is a conceptually significant distinction between the dynamical variables with which particle and wave-like characteristics of matter are quantified, e.g., the variables connected by the Einstein and de Broglie relations. Characteristics like energy and linear momentum can well pertain to individual particles. By contrast, the corresponding quantities of frequency and wavelength—and ultimately, of course, the wave function— although commonly spoken of in the context of single-particle wave packets, actually characterize a hypothetical ensemble of particles all similarly prepared. Although one can in principal measure the mass, charge, or energy of a single electron (held, for example, in an electromagnetic trap), one cannot measure its de Broglie wavelength except by a diffraction or interference experiment employing many such electrons.

And so the original quantum mechanical enigma distills into this: Is the fabric of nature so constructed that the laws of motion are at best statistical, ultimately pertaining to *systems* of particles only and not to individual constituents? Is it physically meaningless even to speak of certain attributes of a particle as objectively real if they can never be simultaneously observed and measured? Although there is little disagreement amongst physicists concerning the conceptual formulation and mathematical procedures of quantum mechanics, the interpretation of the theory and assessments of its fundamentality have elicited over the years a broad range of opinion for which the reader may consult the literature [4].

That electrons behave singly as particles and collectively as waves is indeed mysterious, but, Feynman's remark notwithstanding, this is not the only quantum mystery. Charged particles can do other things that are equally strange—indeed, in some ways even stranger. They can interact with electric and magnetic fields through which they do not pass. They can arrive at detectors in classically inexplicable cluster patterns although emitted apparently randomly from their source. And, once part of a localized system of particles, they exhibit long-range correlations that strongly affect their subsequent self-interference well after the original system has apparently ceased to exist and the constituent particles become widely dispersed. It is to be stressed, of course, that the rhetorical term "mystery" does not refer in any way to the inability of current physical theory to account for the phenomena under discussion, but only to the insufficiency of our ordinary experience (i.e., classical physics) to permit us to imagine some tangible mechanism by which the processes might occur.

1.2. Confined Fields and Electron Interference

It has long been a fundamental proposition of modern physics (the origin of which dates back well before the creation of special relativity to at least the time of Michael Faraday) that material systems interact with one another—not instantaneously at a distance—but causally through the medium of a field. The first, and still the most familiar, implementation of this philosophical perspective of nature was in the area of electromagnetism. The mathematical embodiment of the field theory of electromagnetism is the set of Maxwell equations

$$\nabla \cdot \mathbf{E} = 4\pi\rho, \tag{3a}$$

$$\nabla \cdot \mathbf{B} = 0, \tag{3b}$$

$$\nabla \times \mathbf{E} = -\partial\mathbf{B}/c\,\partial t, \tag{3c}$$

$$\nabla \times \mathbf{B} = \partial\mathbf{E}/c\,\partial t + 4\pi\mathbf{J}/c, \tag{3d}$$

and the Lorentz force law

$$\mathbf{F} = \rho\mathbf{E} + \mathbf{J} \times \mathbf{B}/c, \tag{3e}$$

which, together, are considered to represent completely the classical inter-actions of electric ($\mathbf{E}$) and magnetic ($\mathbf{B}$) fields with each other and with charge (ρ) and current ($\mathbf{J}$) densities. The very expression of the laws of electro-dynamics as differential equations seems to signify that all interactions take place locally, the charged particles being influenced only by electric and magnetic fields in their immediate vicinity. Every well-formulated problem in classical electrodynamics essentially reduces to determining the fields produced by a system of charges (stationary or moving) and reciprocally the forces exerted on these charges by the fields. To facilitate the solution of such a problem, vector and scalar potential fields, $\mathbf{A}$ and ϕ, related to the electric and magnetic fields by the defining expressions

$$\mathbf{E} = -\nabla\phi - \partial\mathbf{A}/c\,\partial t, \tag{4a}$$

$$\mathbf{B} = \nabla \times \mathbf{A}, \tag{4b}$$

are ordinarily introduced. The electromagnetic potentials are not unique, but can be modified by a so-called gauge transformation

$$\mathbf{A} \to \mathbf{A}' = \mathbf{A} + \nabla\Lambda, \tag{5a}$$

$$\phi \to \phi' = \phi - \partial\Lambda/c\,\partial t, \tag{5b}$$

with gauge function Λ which leaves the electromagnetic fields, and hence the Maxwell equations and Lorentz force law, invariant. In order to leave invariant, as well, the quantum mechanical equation of motion of a particle with charge q, the gauge transformation also modifies the wave function

$$\Psi \to \Psi' = \Psi e^{iq\Lambda/\hbar c}. \tag{5c}$$

The transformation function Λ is largely arbitrary although usually required to be a single-valued function in order that the contour integral of gauge-transformed vector potentials leads to the same value of magnetic flux Φ through Stokes's law

$$\oint_C \mathbf{A}\cdot\mathbf{ds} = \iint_S \mathbf{B}\cdot\mathbf{dS} = \Phi, \tag{5d}$$

where C is the contour bordering the open and orientable surface S through which the magnetic field lines pass. The fields $\mathbf{E}$ and $\mathbf{B}$ must themselves be unique for a given configuration of charges and currents, since they are directly related to electromagnetic forces. As a consequence, the classical perspective has been to regard $\mathbf{E}$ and $\mathbf{B}$ as the primary or fundamental fields and $\mathbf{A}$ and ϕ as auxiliary or secondary fields needed for calculational convenience only.

It is of historical interest to note, however, that Maxwell, himself, who introduced these fields in his famous treatise [5], accorded a more physical significance to the vector potential. Having initially termed **A** the "vector-potential of magnetic induction," Maxwell subsequently designated it the "electromagnetic momentum at a point" and interpreted **A** as presenting the "direction and magnitude of the time-integral of the electro-motive intensity which a particle placed at [a point] would experience if the primary current [in one of two interacting circuits] were suddenly stopped." In other words, Maxwell regarded **A** as a measurable quantity related to momentum, a conception that may be found, albeit sharpened by the use of modern terminology, in the contemporary physics literature [6]. Nevertheless, the requirement that physical observables be representable by gauge-invariant expressions underlies a long-standing belief that the electro-magnetic potentials, though intimately related to measurable quantities, are not themselves directly observable.

The theoretical demonstration in 1959 by Y. Aharonov and D. Bohm (to be designated from this point on as AB) [7] that the diffraction of charged particles can be influenced by electromagnetic poentials under conditions where the electromagnetic fields are *null* opened a new chapter in the study of quantum interference phenomena and gave rise to controversial issues, both theoretical and experimental, that have spanned some thirty years. Obviously, one of the principal questions raised by AB concerned whether or not the vector and scalar potentials were more fundamental than the electric and magnetic fields. AB argued that there were. The quantum implications exposed by the AB paper had actually been revealed some ten years earlier by Ehrenberg and Siday [8] who were investigating the refractive index of electrons in an electron microscope. That paper, however, was apparently not widely read, and it is through AB that the physics community came to recognize the extraordinary consequences of electro-magnetic potentials in quantum mechanics.

The basic mathematical relation underlying the AB effect seems to have been known at least as far back as the early 1930s by P.A.M. Dirac in a celebrated study of quantized singularities (magnetic monopoles) [9]. Expressed more generally to embrace both vector and scalar potentials, it is this. If Ψ_0 is the solution to the quantum equations of motion (e.g., the Dirac or Schrödinger equation) in the absence of electromagnetic interactions, then the corresponding wave function Ψ of a charged particle in the presence of a time-independent vector potential field and a spatially uniform scalar potential field takes the form

$$\Psi(\mathbf{r}, t) = \Psi_0(\mathbf{r}, t)e^{iS(\mathbf{r}, t)}, \tag{6a}$$

where the phase $S(\mathbf{r}, t)$ is given by

$$S(\mathbf{r}, t) = (q/\hbar c)\left[\int_{\mathbf{r}_o}^{\mathbf{r}} \mathbf{A} \cdot \mathbf{ds} - \int_{t_o}^{t} c\phi\, dt\right]. \tag{6b}$$

The integration in the phase is over an arbitrary space–time path between some point of origin $(t_o, \mathbf{r}_o)$ and destination $(t, \mathbf{r})$.

The above result is demonstrable by direct substitution of the wave function (6a) into the wave equation

$$H\Psi = i\hbar\, \partial\Psi/\partial t, \tag{7a}$$

where the Hamiltonian H is constructed from the field-free Hamiltonian H_0 by replacing the canonical linear momentum $\mathbf{p}$ with $\mathbf{p} - q\mathbf{A}/c$ and adding $q\phi$ to the potential energy. The resulting wave equation is then of the form

$$H'\Psi_0 = i\hbar\, \partial\Psi_0\, \partial t, \tag{7b}$$

where the transformed Hamiltonian H'

$$H' = U^\dagger H U - i\hbar U^\dagger\, \partial U/\partial t \tag{7c}$$

reduces to the field-free Hamiltonian H_0 when $\partial\mathbf{A}/\partial t$ and $\nabla\phi$ both vanish.

It is worth noting that the operator $\mathbf{p} - q\mathbf{A}/c$, which corresponds to the kinetic linear momentum $\mathbf{P}$ (equal to $m\mathbf{v}$ for a nonrelativistic classical particle), gives rise to a gauge-independent expectation value, whereas the expectation value of $\mathbf{p}$, the dynamical variable entering into the quantum commutation relation

$$[\mathbf{r}, \mathbf{p}] = i\hbar\mathbf{1}, \tag{8}$$

and serving as the generator of spatial translations, is gauge dependent. (In (8), $\mathbf{1}$ is the unit dyad or second-order unit tensor.) Although both operators are Hermitian, $\mathbf{P}$ is considered a dynamical observable, but $\mathbf{p}$ is not. That the kinetic and canonical linear momenta are not equivalent in the presence of electromagnetic potentials is not unique to quantum mechanics, but is known as well, although is less consequential, in classical mechanics [10].

One type of current configuration that ideally gives rise to a null magnetic field in a spatial region where the vector potential does not vanish is that of an infinitely long axial coil or solenoid. Within the solenoid the magnetic field $\mathbf{B}$ is parallel to the symmetry axis with a strength and orientation, respectively, determined by the magnitude and sense of the current circulation through the windings. The magnetic flux Φ within an infinitely long coil of radius R is simply the product of the field strength and the cross section area πR^2. Outside the solenoid the magnetic field is ideally null. The vector potential, however, forms cylindrical equipotential surfaces in both regions of space (with a sense of circulation opposite that of the electron current). In the Coulomb gauge—the gauge for which $\nabla \cdot \mathbf{A} = 0$—the tangential (and only) component of $\mathbf{A}$ is a function of the radial coordinate r given by

$$A(r) = \begin{cases} \Phi/2\pi r & (r \geq R), \\ \Phi r/2\pi R^2 & (r \leq R). \end{cases} \tag{9}$$

For any real solenoid, of course, there is a return field in the exterior region, but the magnitude of this field diminishes with increasing ratio of length to radius [11]. Besides the ideally infinite solenoid, a toroidal current configuration formed by joining the two ends of a finite solenoid accomplishes the same task of producing a confined magnetic field (although the expressions for the resulting vector potential field are not as simple).

Feynman has pointed out the dramatic consequence of the seemingly innocuous relation (6a, b) in the context of a two-slit particle interference experiment [12] with an ideal solenoid placed behind the diffraction screen and between the slits (the long axis parallel to the slits). The experiment is conceptually simpler than the configuration orginally analyzed by AB which involved scattering of charged particles directly incident upon the solenoid. It is the Feynman version of the AB effect (actually first described in the paper by Ehrenberg and Siday) that has generally found its way into physics textbooks [13].

Reduced to its essentials, as schematically shown in Figure 1.5, there are two types of classically indistinguishable pathways by which an incident particle can propagate from its source S to the detector D: by going clockwise above the solenoid (Path I) or counterclockwise below the solenoid (Path II). It is assumed that the particles never penetrate the solenoid, and that in the absence of a current through the solenoid there is complete symmetry above and below the forward direction of the beam. Since there is a vector potential (but no scalar potential) in the space accessible to the electrons, the probability amplitude for each pathway, when the solenoid interior

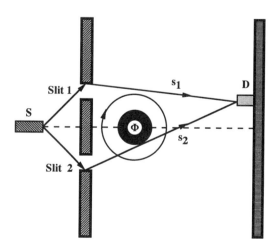

Figure 1.5. Schematic diagram of the two-slit Aharonov–Bohm (AB) effect. A coherently split electron wave front, issuing from source S, passes to one side or the other of a region of confined magnetic flux Φ and is recombined at the detector D. The resulting interference pattern is influenced by the magnetic field through which the electrons do not pass.

contains an axial magnetic field, takes the form of (6a) with ϕ equal 0 in the phase (6b). Hence the total probability amplitude for a particle to be received at D can be written as

$$\Psi(D) = e^{i\Delta}[1 + e^{i(\delta_0 + \delta)}]/\sqrt{2}, \tag{10a}$$

where Δ is an inconsequential global phase, δ_0 is the relative phase (dependent upon optical path length difference) when the magnetic field is zero, and

$$\delta = (q/\hbar c)\left(\int_{\text{Path I}} \mathbf{A} \cdot \mathbf{ds} - \int_{\text{Path II}} \mathbf{A} \cdot \mathbf{ds}\right) = (q/\hbar c)\oint_C \mathbf{A} \cdot \mathbf{ds} = 2\pi\Phi/\Phi_0 \tag{10b}$$

is the relative phase contributed by the confined magnetic field. The ratio Φ/Φ_0 expresses the magnetic flux through the solenoid in units of the quantum of flux or fluxon, which for an electron is

$$\Phi_0 \equiv hc/e = 4.14 \times 10^{-7} \text{ Gauss cm}^2, \tag{10c}$$

The contour C marking the integral in (10b) is a closed path between S and D circumscribing the solenoid in a clockwise sense. If the magnetic field within the coil is directed into the plane of the paper, then the external vector potential field circulates in the same sense as the contour C, and the sign of the magnetic flux is positive. It follows from relations (10a, b) that the probability of receiving a particle at D is

$$P(D) = (\tfrac{1}{2})[1 + \cos(\delta_0(D) + 2\pi\Phi/\Phi_0)]. \tag{10d}$$

The physical content of the above relation is that a magnetic field, from which region the charged particles are totally excluded, can influence the spatial distribution of the particles. In fact, the influence can be strong. For D located in the forward direction ($\delta_0 = 0$), there is constructive interference and thus 100% probability of receiving a particle in the absence of the magnetic field ($\Phi = 0$). However, with a solenoidal magnetic field of such strength that $2\pi\Phi/\Phi_0$ is an integral multiple of π radians, the probability $P(D)$ is zero; the isolated magnetic field has converted a bright fringe (maximum) into a dark fringe (minimum) in the resulting interference pattern. What is one to make of such a phenomenon whereby particles can be seemingly displaced from their "intended" direction without the agency of an external force?

To some, the answer has been that the AB effect does not physically exist, that it is mere mathematical self-delusion [14]. Although the details of the underlying reasoning must be left to the original literature, the core of the argument concerns the arbitrariness of gauge transformations. In short, one can find a gauge, the proponents claim, for which the vector potential vanishes entirely from the equation of motion which thereby describes a system in an environment free of any electromagnetic influence. The argument is fallacious, however, for the proposed transformation removes not only the vector potential in the space accessible to the particles, but alters the magnetic field in the interior of the solenoid as well. As pointed out earlier, no gauge transformation is admissible that changes the physical configuration of the electromagnetic fields. Moreover, aside from the theoretical inconsistency of the argument, there is substantial experimental evidence in support of the existence of the AB effect.

To others, who accept the experimental confirmations of the predicted fringe shifts, the answer has been that the AB effect is essentially a consequence of, or is equivalent to, the classical Lorentz force. Feynman, for example, in his *Lectures*, described the action of the solenoid as essentially equivalent to that of a magnetic strip placed behind the two slits. A magnetic field, however, would displace the *forward* beam direction, i.e., the *center* of the single-slit diffraction pattern. This is not what should occur in the AB effect. Analyses of model configurations [15] show that, in the absence of a local magnetic field (a magnetic field in the space through which the charged particles propagate) the pattern of interference fringes is displaced within the enveloping diffraction pattern which is itself unaffected. Were this not the case, the AB effect would conflict with the Bohr correspondence principle which requires quantum mechanics to give results compatible with classical mechanics in a domain for which both theories are valid. In this regard it is interesting to note that there is usually more than one way to extract the classical limit of a quantum calculation—e.g., one can let Planck's constant h approach zero, or let some quantum number approach infinity—and the different ways are not always equivalent [16]. The AB effect, as described so far, occurs with unbound particles; the classical limit is suitably taken by letting $h \to 0$ in which case the spatial periodicity of the fringe pattern becomes infinitesimally small and the central diffraction spot undeflected.

To the majority of physicists concerned with this fundamental issue, there remains the final option of accepting the AB effect for what it appears to be: a force-free interaction with either a local vector potential field or a nonlocal magnetic field. The issue of locality versus nonlocality is actually deceptive, the two points of view being equivalent in representing the AB effect as an intrinsically nonlocal physical phenomenon. Although it is the case that charged particles interact directly with the vector potential field at their instantaneous positions, this local interaction in itself is not sufficient to produce the AB effect. The allowed paths of the particles must circumscribe a region of space within which the magnetic field is confined and from which the particles are excluded. The AB effect, therefore, reflects the global geometry of the space accessible to the particles. In the simple two-slit configuration represented in Figure 1.5, the ambient space has the topology of a doughnut.

What, in view of the AB effect, is a reasonable posture to take regarding the fundamentality of electromagnetic fields and potentials? One widely accepted interpretation (although perhaps not so widely known throughout the physics community as a whole) has been articulated by Wu and

Yang [17], according to whom an intrinsic and complete description of electromagnetism is provided by the nonintegrable (i.e., path-dependent) phase factor $\exp(iS(\mathbf{r}, t))$ in (6a). It is the phase *factor*, and not the phase (6b) alone, that is physically meaningful, because the phase (which manifests the arbitrariness of the potentials) contains more information than is determinable by measurement. Conversely, the fields $\mathbf{E}$ and $\mathbf{B}$ contain less information than is measurable and therefore provide an incomplete description of electromagnetism when quantum processes are taken into account. Note that these are *classical* fields; the quantum processes refer here only to matter.

The experimental confirmation of the AB effect, upon which the post-Maxwellian interpretation of electrodynamics ultimately rests, has never been a simple matter, in part because of the extreme difficulty of producing a magnetic field with no leakage into the spatial region of the particles. In the earliest AB experiments confined magnetic fields were produced by either tiny ferromagnetic filaments (or whiskers) or microscopic solenoids. Although the results were in accord with theoretical predictions derived from the nonintegrable phase factor, the fact that the ideal conditions of the AB effect had not been met allowed critics to point to the classical Lorentz force as the causative agent for any fringe shifts. While a complete survey of AB experiments must be left to the literature [18], it is relevant to mention here two experiments, one of the earliest and one of the most recent, of particular conceptual interest.

In the 1962 experiment of Bayh [19] the AB effect was recorded on photographic film by a 40 kV electron beam split and recombined by a system of three electrostatic biprisms. Between the first and second biprism, at a location where the extent of separation of the electron beam is greatest, was inserted a tiny tungsten coil about 5 mm in length and less than 20 μm in diameter to serve as the AB solenoid. Detailed calculations of the spatial variation of the coil magnetic field (based on an equation of Buchholz [20]) indicated that for a coil of 20 μm diameter and pitch of 6 μm the radial component of the magnetic field in the midplane of the coil and at a distance of 10 μm from the coil windings was weaker than the interior axial field by approximately 2×10^{-5}. Since the components of the split electron beam could be separated by 50–60 μm without exceeding the coherence condition which had to be met for interference to occur, Bayh concluded that the nonideal effects of the coil should be negligibly small.

The AB phase shift was demonstrated by fastening the film, upon which the interference pattern was to be recorded, to a small electric motor and advancing the film at a rate proportional to the rate of increase of current through the windings of the coil. The film was shielded except for a 0.5 mm wide slit oriented perpendicular to the interference fringes so that each narrow horizontal section through the interference pattern corresponded to a well-defined value of magnetic flux through the solenoid. The resulting interference pattern, shown in Figure 1.6, showed the continuous lateral

Figure 1.6. Demonstration of electron-wave phase shift in the presence of a vector potential field (AB effect) in the Bayh experiment. The magnetic field is held constant in the upper and lower third of the figure; in the middle third the variation in interference fringes follows a linear variation in magnetic field strength. (Adapted from Bayh [19].)

displacement of the fringes (for a total distance of roughly four times the fringe spacing) *within* the enveloping pattern (produced by diffraction around the biprism filament) which remained unchanged despite the variation in vector potential and magnetic fields.

What is especially interesting about this experiment is that, as a result of the time-varying magnetic flux, the fringe shifting may also be accounted for by an apparently purely classical argument, one based on Faraday's law of induction

$$\mathscr{E}(t) = -\partial\Phi(t)/c\,\partial t. \tag{11a}$$

Here $\mathscr{E}(t)$ is the electromotive force

$$\mathscr{E}(t) = \oint_C \mathbf{E}(t)\cdot \mathbf{ds}, \tag{11b}$$

induced around the solenoid in the exterior region through which the electrons pass. The induced electric field does work on the electrons and thereby engenders a relative phase

$$\theta = (1/\hbar)\int e\mathscr{E}(t)\,dt = (e/\hbar c)\int d\Phi = 2\pi\Phi/\Phi_0, \tag{11c}$$

which is exactly what one would have expected on the basis of the force-free AB effect.

The difficulty with this classical interpretation, however, is apparent once one realizes again that it is not a succession of classical wave fronts, but rather discrete and uncorrelated electrons that passed through the interferometer. With an energy of 40 kV, and hence a velocity of ≈ 0.38c, an electron propagated from source to film—a distance of the order of 1 m—in about 0.01 µs. Although Bayh did not specify the rate at which the current increased through the solenoid windings, it was undoubtedly over a much longer time interval. Consequently, the interference pattern was created over a relatively long period of time by the arrival of a large number of independent electrons, each one sampling an effectively instantaneous value of the local vector potential field. Nevertheless, the fact that the space accessible to the electrons was not entirely devoid of a force field was a potential source of criticism.

More recently, researchers at the Hitachi Advanced Research Laboratory produced the AB effect under conditions more closely duplicating the requirements of an ideal force-free environment than had been attained previously [21]. The salient novel feature of the Hitachi experiment, which, like antecedent experiments, employed a Möllenstedt biprism to split the beam of an electron microscope, was the use of a microscopic (≈ 4 µm diameter) toroidal ferromagnet, in place of a "whisker" or solenoid, as the source of a confined magnetic field. Ideally, the magnetic field lines circulate within the toroid about the C_∞ symmetry axis (i.e., the axis perpendicular to the plane of the toroid). However, to guard against possible leakage of the magnetic field, the toroid was covered completely with a superconductive layer of niobium. When brought below the critical temperature $T_c = 9.2$ K, the niobium underwent a transition to the superconducting state and expelled magnetic flux from its interior, and therefore *into* the permalloy toroid, by means of the Meissner effect [22]. An additional layer of (nonsuperconductive) copper further helped reduce penetration of the 150 kV electrons into the toroid.

In the absence of a magnetic toroid, the split electron beam gives rise to the standard pattern of parallel fringes in the observation plane. With the magnet in place (above the biprism), theoretical analysis predicts an AB phase shift between components of an electron wave packet propagating through the central hole of the toroid compared with passage around the outer periphery. When the experiment was performed with an unshielded toroidal magnet, one could see in the resulting interferogram the continuous displacement of a light or dark fringe from the exterior region, across the body of the annulus, into the region of the hole as shown in Figure 1.7. Of course, without the shielding layers electrons can penetrate the magnet, and critics could again attribute phase shifts to classical effects of the Lorentz force.

The use of toroids with a superconducting outer layer, however, had an unanticipated and potentially adverse "side-effect," namely, that the magnetic flux trapped within the annulus became quantized in units of one-half a fluxon. This quantization condition

$$\Phi = nhc/2e = n\Phi_0/2, \tag{12}$$

where n is an integer, pertains to *flux*, not fields, and should not be confused

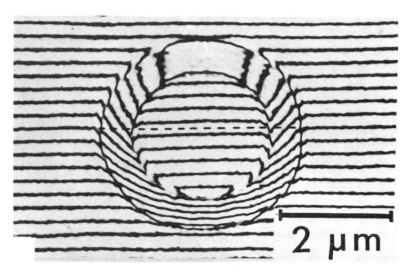

Figure 1.7. Observation of the AB effect with an unshielded toroidal ferromagnet. The fringes in the electron interference pattern, continuous across the outline of the toroid within which the magnetic field is confined, are uniform, parallel, and shifted within the zero-field region of the hole. (Courtesy of A. Tonomura, Hitachi ARL.)

with the quantization of the electromagnetic field—i.e., the description of electromagnetic waves in terms of photons—which plays no role in the present system. Rather, the flux quantization condition is a consequence of the fact that the charge carrier within the superconductor is not a single electron, but a (Cooper) pair of electrons, and that the wave function of this pair is macroscopically coherent around the annulus. As a result of flux

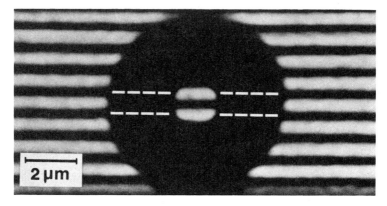

Figure 1.8. AB effect experiment performed with a toroidal ferromagnet shielded by an outer superconducting layer. The fringe reversal between the external region and the region of the hole represents an AB phase shift of 180°. (Courtesy of A. Tonomura, Hitachi ARL.)

quantization within the toroid, the magnetic phase shift in (10d) becomes $2\pi\Phi/\Phi_0 = n\pi$ and is either 0 (mod 2π) or π (mod 2π) according to whether n is an even or odd integer. In the first case the AB effect leads to no observable outcome. In the second case, however, there is a complete fringe reversal from maximum to minimum between the exterior region and the central hole. Experimentally, the fabricated toroids used in the Hitachi experiment produced a range of discrete flux values, both odd and even. The observation of the predicted 180° phase reversal shown in Figure 1.8, provides the strongest evidence to date that the spatial distribution of charged particles can be strongly altered by a magnetic field which the particles never encounter directly.

References

[1] R.P. Feynman, R.B. Leighton, and M. Sands, *The Feynman Lectures on Physics*, Vol. 3 (Addison-Wesley, Reading, MA, 1965), p. 1-1.

[2] A. Tonomura, J. Endo, T. Matsuda, and T. Kawasaki, Demonstration of Single-Electron Build-Up of an Interference Pattern, *Amer. J. Phys.*, **57**, 117–120 (1989).

[3] M. Born and E. Wolf, *Principles of Optics*, 4th edn. (Pergamon, Oxford, 1970), pp. 267–268.

[4] See, for example, M. Jammer, *The Conceptual Development of Quantum Mechanics* (McGraw-Hill, New York, 1966) and *The Philosophy of Quantum Mechanics* (Wiley, New York, 1974).

[5] J.C. Maxwell, *A Treatise on Electricity and Magnetism*, 3rd edn., Vol. 2 (Dover, New York, 1954) pp. 232, 257.

[6] E.J. Konopinski, What the Electromagnetic Vector Potential Describes, *Amer. J. Phys.*, **46**, 499–502 (1978).

[7] Y. Aharonov and D. Bohm, Significance of Electromagnetic Potentials in the Quantum Theory, *Phys. Rev.*, **115**, 485–491 (1959).

[8] W. Ehrenberg and R. E. Siday, The Refractive Index in Electron Optics and the Principles of Dynamics, *Proc. Phys. Soc. London*, **B62**, 8–21 (1949).

[9] P.A.M. Dirac, Quantised Singularities in the Electromagnetic Field, *Proc. Roy. Soc.* **A133**, 60–72 (1931).

[10] See, for example, H. Goldstein, *Classical Mechanics*, 2nd edn. (Addison-Wesley, Reading, MA, 1980) pp. 346, 361.

[11] E.M. Purcell, *Electricity and Magnetism*, 2nd edn. (McGraw-Hill, New York, 1985) pp. 226–231.

[12] R.P. Feynman, R.B. Leighton, and M. Sands, *The Feynman Lectures on Physics*, Vol. 2 (Addison-Wesley, Reading, MA, 1965), pp. 15.12–15.14.

[13] See, for example, G. Baym, *Lectures on Quantum Mechanics* (Benjamin, New York, 1969), pp. 77–79; J.J. Sakurai, *Advanced Quantum Mechanics* (Addison-Wesley, Reading, MA, 1967), pp. 16–18.

[14] P. Bocchieri and A. Loinger, Nonexistence of the Aharonov–Bohm Effect, *Il Nuovo Cimento*, **47A**, 475–482 (1978); Nonexistence of the Aharonov–Bohm Effect.—II: Discussion of the Experiments, *Il Nuovo Cimento*, **51A**, 1–17 (1979).

[15] S. Olariu and I. Popescu, The Quantum Effects of Electromagnetic Fluxes, *Rev. Mod. Phys.*, **57**, 339–436 (1985).

[16] R.L. Liboff, The Correspondence Principle Revisited, *Physics Today*, **37**, No. 2, 50–55 (1984).

[17] T.T. Wu and C.N. Yang, Concept of Nonintegrable Phase Factors and Global Formulation of Gauge Fields, *Phys. Rev. D*, **12**, 3845–3857 (1975).

[18] Detailed reviews of AB experiments are provided in [15] and in M. Peshkin and A. Tonomura, *The Aharonov–Bohm Effect*, Lecture Notes in Physics, Vol. 340 (Springer-Verlag, Berlin, 1989).

[19] W. Bayh, Messung der kontinuierlichen Phasenschiebung von Elektronenwellen im kraftfeldfreien Raum durch das magnetische Vektorpotential einer Wolfram-Wendel (Measurement of the Continuous Phase Shift of Electron Waves in Force-Free Space by the Magnetic Vector Potential of a Tungsten Solenoid), *Z. Physik*, **169**, 492–510 (1962).

[20] H. Buchholz, *Elektrische und Magnetische Potentialfelder* (Springer-Verlag, Berlin, 1957), Chapter 6, pp. 112, 270.

[21] N. Osakabe, T. Matsuda, T. Kawasaki, J. Endo, A. Tonomura, S. Yano, and H. Yamada, Experimental Confirmation of Aharonov–Bohm Effect Using a Toroidal Magnetic Field Confined by a Superconductor, *Phys. Rev. A*, **34**, 815–822 (1986).

[22] For a discussion of the Meissner effect, see, for example, D.R. Tilley and J. Tilley, *Superfluidity and Superconductivity*, 2nd edn. (Adam Hilger, Bristol, 1986), pp. 145–150.

CHAPTER 2

Around and Around: The Rotating Electron in Electromagnetic Fields

The Earth, the Air, the Fire, the Water
Return, Return, Return, Return
The Earth, the Air, the Fire, the Water
Return, Return, Return, Return

Round compiled by Libana
Source unknown

2.1. Broken Symmetry of the Charged Planar Rotator

In the atomic model introduced by Niels Bohr in 1913 the electron was presumed to follow a circular orbit about the atomic nucleus. Although the nonrelativistic quantum theory of the hydrogen atom later confirmed the Bohr atom energy spectrum (in the absence of spin-related relativistic interactions), it also showed that the concept of orbits was in general no longer tenable. The bound electron could have any possible separation from the nucleus according to a distribution function determined from the radial wave equation.

Nevertheless, the two-dimensional charged rotator—superficially a Bohr atom without nucleus in which the electron is confined by nonspecified forces to a circular space of radius R—is an interesting quantum system to consider (Figure 2.1). The orbiting particle does not, of course, follow a trajectory just like that of the electron in the Bohr atom. A more appropriate, albeit approximate, real-world counterpart to this model system is a pi electron in a conjugated hydrocarbon ring [1]. Indeed, the planar rotator (with concomitant application of the Pauli exclusion principle) yields a moderately successful prediction of the center of gravity of the rotational Raman spectrum of the hexagonal molecule benzene, the archetype of conjugated aromatic molecules (Figure 2.2) [2].

The field-free planar rotator is a quantum system that, like the particle in a rectangular box with infinite walls, is trivially easy to analyze. The spectral degeneracy can be simply predicted on the basis of circular symmetry.

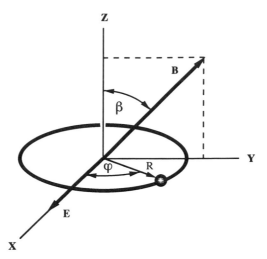

Figure 2.1. Field orientations for the Stark and Zeeman effects of the two-dimensional charged rotator.

Since the two possible classical motions of clockwise and counterclockwise rotation at the same angular frequency are energetically equivalent, all rotationally excited levels are doubly degenerate. The ground state, in which the angular momentum, and therefore the rotational energy, is zero, is nondegenerate. It is worth noting that there is no zero-point energy as, for example, in the case of the harmonic oscillator, since the electron is not

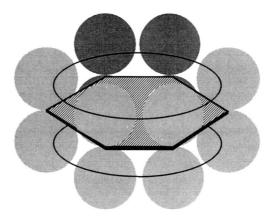

Figure 2.2. The benzene molecule, C_6H_6. Each vertex represents a carbon atom; the hydrogen atoms (one bonded to each carbon) are not shown. Sigma bonds, strong linkages of carbon $2s$ electrons directed between adjacent atoms in the molecular plane, hold the framework together. Weaker pi bonds, linkages of carbon $2p$ electrons oriented normal to the plane, provide a ring-like network (indicated schematically by the rings) over which the pi electrons are delocalized.

bound in a potential well but occupies the entire space available to it. This space happens to be a one-dimensional space of constant curvature. From the form of the Hamiltonian for a particle with mass m confined to a ring of radius R

$$H_0 = L_z^2/2mR^2, \tag{1a}$$

it follows immediately that the state vectors can be labeled by an integer quantum number μ (with $\mu = 0$, ± 1, ± 2, etc.) designating the eigenvalues $\mu\hbar$ of L_z, the component of angular momentum along the rotation axis (and, in fact, the only component of angular momentum). The energy eigenvalues are then

$$E_\mu = E_{-\mu} = h^2\mu^2/8\pi^2mR^2, \tag{1b}$$

with associated normalized wave functions

$$\langle \varphi | \mu \rangle \equiv \psi_\mu(\varphi) = e^{i\mu\varphi}/\sqrt{2\pi}, \tag{1c}$$

that satisfy the boundary condition

$$\psi_\mu(\varphi + 2\pi) = \psi_\mu(\varphi) \tag{1d}$$

required for single-valued functions. The azimuthal angle φ is measured in a positive sense from some arbitrarily chosen axis (which is usually the x-axis of a Cartesian coordinate system). The states represented by wave function (1c) are also eigenstates of the parity operator Π which reflects the particle through the origin ($\varphi \to \varphi + \pi$) with eigenvalues $e^{i\mu\pi} = (-1)^\mu$. Thus, states of even or odd μ, respectively, have even or odd parity. It should also be noted, in anticipation of an important point to be discussed shortly, that any linear combination of states of fixed $|\mu|$ is also a valid solution of the Schrödinger equation.

With the introduction of external electric or magnetic fields or electromagnetic potentials, the problem of the charged planar rotator ceases to be a trivial one [3, 4]. Although the system may still seem simple, this simplicity is somewhat deceiving. As one of few quantum mechanical systems involving electromagnetic interactions amenable to exact solution, there is much of conceptual importance that one can learn from studying it. For one thing, the Schrödinger equations for both the Stark and Zeeman effects lead to differential equations of a type not frequently encountered in quantum mechanics: one with periodic coefficients and a three-term recursion relation. Such an equation is not expressible in terms of the generators of a particular symmetry group (known as the SU(1, 1) dynamical group [5]) as are those of a number of commonly encountered solvable quantum systems, as, for example, the Coulomb and harmonic oscillator potentials. Moreover, the energy spectrum of the resulting Stark and Zeeman states exhibits an unusual broken degeneracy pattern. Indeed, in at least one standard reference, it had not been suspected that the degeneracy is broken at all [6].

The planar rotator in a vector potential field in the *absence* of all electric and magnetic fields—in other words, the bound-state counterpart to the AB effect—is a particularly interesting model system for examining another problematic aspect of the nonclassical effects of potentials. At issue is not the flux dependence of a diffraction pattern or scattering cross section, but rather the angular momentum spectrum, rotational behavior, and statistics of particles in rotational motion about tubes of magnetic flux [7, 8]. Some have proposed that a charged particle orbiting a magnetic flux tube can have a flux-dependent—and therefore arbitrary—angular momentum, in which case the wave function would acquire a flux-dependent rotational phase factor. An ensemble of such systems would follow statistics that interpolate between boson and fermion statistics [9]. Others have maintained instead that the orbital angular momentum spectrum remains integer-valued, and the wave function of the rotated particle is unaffected by the flux [10]. A collection of composite rotator-flux systems would then obey the traditional quantum statistics determined by the spin of the particle. Thus, as with earlier controversies of the AB effect, these, too, raise questions of fundamental import that transcend in scope the particularities of the individual systems under study. These questions concern, for example, the role, spectrum, and observability of various gauge invariant and noninvariant dynamical quantities in quantum mechanics.

In the following sections we will examine the quantum behavior of a two-dimensional rotator in the presence of electric and magnetic fields and the interaction of a rotator with a vector potential in an otherwise field-free space.

2.2. The Two-Dimensional Rotator in an Electric Field

Let us start with the Stark effect [11]. Without loss of generality consider the electron to be confined to a circle of radius R in the x–y plane centered on the z axis. Since only components of an electric field $\mathbf{E}$ within the orbital plane can influence the state of motion of the electron, we orient $\mathbf{E}$ along the x axis as shown in Figure 2.1. The Hamiltonian characterizing the total rotational and electric energy of a two-dimensional rotator with electric dipole moment

$$\boldsymbol{\mu_E} = e\mathbf{r} = eR(\cos\varphi, \sin\varphi, 0) \tag{2}$$

$(e = -|e|)$ is then

$$H = H_0 + H_E = L_z^2/2mR^2 - eEx, \tag{3a}$$

or, in a coordinate representation,

$$H = -\varepsilon(\partial^2/\partial\varphi^2 - \lambda\cos\varphi) \tag{3b}$$

with energy parameter

$$\varepsilon = h^2/8\pi^2mR^2, \tag{3c}$$

and electric interaction parameter (which is positive and dimensionless)

$$\lambda = 8\pi^2m|e|R^3E/h^2. \tag{3d}$$

One may be tempted on the basis of common textbook examples of perturbation theory to predict that the perturbed energy spectrum of the planar rotator would display a quadratic Stark effect in the nondegenerate ground state (i.e., energy increases as E^2), and a linear Stark effect in the doubly degenerate excited states (energy of one state increases and that of the other decreases as E^1). This description of the degeneracy breaking, however, is not correct. The matrix elements of H_E between the unperturbed states are readily calculated to be

$$\langle\mu|H_E|\mu'\rangle = -\tfrac{1}{2}eER[\delta_{\mu,\mu'+1} + \delta_{\mu,\mu'-1}]. \tag{4}$$

As expected, the first-order contribution to the ground state vanishes since H_E has no diagonal matrix elements. It also follows from relation (4), perhaps surprisingly, that there can be no first-order splitting of the degenerate pairs $|\pm\mu\rangle$. For splitting to occur, as in the case of the degenerate 2S, 2P($m = 0$) states of hydrogen (in the absence of spin), the degenerate states must be coupled by a matrix element of the interaction. In the case of hydrogen, the element $\langle 2P(m = 0)|z|2S\rangle$ is not zero. In the case of the planar rotator, however, the matrix element $\langle -\mu|x|\mu\rangle$ vanishes unless $-\mu = \mu \pm 1$, leading to $\mu = \pm\tfrac{1}{2}$. Since there are no half-integer eigenvalues in the spectrum of the orbital angular momentum L_z, breaking of the excited state degeneracy must occur at least quadratically in E.

Additional physical insight can be gained by considering the relationships between parity and reflection symmetry and the electric dipole moment. The ground states of both the planar rotator and the hydrogen atom are parity eigenstates. Hence, the expectation value of the electric dipole moment in these states vanishes identically since

$$\langle 0|\boldsymbol{\mu}_E|0\rangle = \langle 0\Pi|\Pi\boldsymbol{\mu}_E|\Pi 0\rangle = \langle 0|-\boldsymbol{\mu}_E|0\rangle = 0, \tag{5}$$

where $\Pi^2 = \Pi$. The external electric field can induce a dipole moment in the ground state proportional to E, where the proportionality constant is the polarizability α. The induced moment then couples to the field giving an electric energy proportional to E^2. Classically, this energy is given by $-\alpha E^2/2$.

For degenerate states a different mechanism is possible. Although two degenerate eigenstates of the Hamiltonian may each be a parity eigenstate, linear combinations of these states need not be. Thus, for example, the hydrogen 2S and 2P($m = 0$) states have parity eigenvalues $+1$, -1, respectively, but the degenerate linear combination $[|2S\rangle \pm |2P(m = 0)\rangle]/\sqrt{2}$ are obviously not parity eigenvectors. They form field-free degenerate eigenstates with a nonvanishing mean electric dipole moment. The external field can couple directly to these dipole moments, and the energy acquired by their alignment is therefore linear in E.

For the two-dimensional rotator the comparable linear combinations $|\mu\pm\rangle = [|\mu\rangle \pm |-\mu\rangle]/\sqrt{2}$ are also not parity eigenstates. However, unlike the hydrogenic states, they do not represent zero-field states with a non-vanishing electric dipole moment. The reason for this is traceable to the reflection symmetry of the Hamiltonian (3a, b). Under the operations Π_x

and Π_y for reflection across the x and y axes, the angular coordinate φ undergoes the changes $(\varphi \rightarrow -\varphi)$, and $(\varphi \rightarrow \pi - \varphi)$, respectively, and

$$\Pi_x |\mu\pm\rangle = \pm|\mu\pm\rangle, \tag{6a}$$

$$\Pi_y |\mu\pm\rangle = \pm e^{i\mu\pi}|\mu\pm\rangle. \tag{6b}$$

It is then straightforward to show by steps analogous to those in (5) that

$$\langle \mu\pm|x|\mu\pm\rangle = \langle \mu\pm|x|\mu\mp\rangle = 0. \tag{7}$$

Since the $|\mu\pm\rangle$ are linearly independent states, there is no possible linear combination of them that can lead to a nonvanishing electric dipole moment. Unlike the case of the hydrogen atom, therefore, the symmetry of the planar rotator precludes a linear Stark effect. It is worth remarking at this point that the relations summarized in (7) are precisely the mathematical conditions for the failure of first-order degenerate perturbation theory to lift the degeneracy of a pair of states.

Let us examine more quantitatively the application of perturbation theory to this system. The states are to be labeled as $|\eta_{\mu\pi}\rangle$ where $\eta^2 \rightarrow \mu^2$ for $\lambda \rightarrow 0$ (vanishing electric field) and $\pi = \pm$ is the eigenvalue under the reflection Π_x. The energy eigenvalues can be written in the form

$$E_{\mu\pi} = \varepsilon\eta_{\mu\pi}^2, \tag{8}$$

and are therefore completely specified by the dimensionless functions $\eta_{\mu\pi}(\lambda)$.

Upon application of standard second-order perturbation theory [12], one obtains

$$\eta_{0+} = -\lambda^2/2 \tag{9a}$$

for the ground state. Only the states with $\mu = \pm 1$ contribute to the infinite summation over unperturbed eigenstates. In view of the preceding discussion, the ground-state polarizability of the two-dimensional rotator is $\alpha_0 = 2R^4/a_0$, where $a_0 = \hbar^2/me^2$ is the Bohr radius; a planar rotator the size of a Bohr atom has twice the polarizability.

Applying second-order *degenerate* perturbation theory to the excited states requires solving the secular equation of a 2×2 matrix where each of the elements is an infinite summation over the unperturbed eigenstates; the two solutions of the resulting quadratic equation yield the second-order correction to the energy of the two degenerate states. For the $\mu = \pm 1$ states one obtains

$$\eta_{1+} = 1 + 5\lambda^2/12, \tag{9b}$$

$$\eta_{1-} = 1 - \lambda^2/12, \tag{9c}$$

which leads to a level separation of $\lambda^2/2$. For all higher states, $|\mu| > 1$, the energy to order E^2 is determined from the equation

$$\eta_{\mu\pm}^2 = \mu^2 + \frac{\lambda^2/2}{4\mu^2 - 1}, \tag{9d}$$

which is a function of the square of μ. It may seem surprising that to second order in E the degeneracy of only the $\mu = \pm 1$ states is lifted. I will comment on this matter in greater detail shortly.

To remove the degeneracy of the states with $|\mu| > 1$ one can try degenerate perturbation theory in still higher orders, but the calculations rapidly become tedious. It is better, instead, to reconsider the entire Schrödinger equation and cast it in a form such that the properties of the exact solution may be most readily obtained. Substitution into the Hamiltonian (3b) of

$$\varphi = 2\theta,$$

$$a = 4\mu^2, \tag{10}$$

$$q = 2\lambda,$$

brings the Schrödinger equation into the form

$$d^2\psi_\mu/d\theta^2 + (a - 2q\cos 2\theta)\psi_\mu(\theta) = 0, \tag{11}$$

recognizable to mathematicians as the canonical form of Mathieu's equation [13].

The Mathieu equation has its origin in investigations of vibrating systems with elliptical boundary conditions and is not uncommonly encountered in wave propagation problems in electromagnetism and acoustics, as well as electron propagation through crystal lattices [14]. Although the above linear differential equation with periodic coefficients is invariant under the transformation $\theta \to \theta + \pi$, the solutions need not be, a point of interest in view of the significance to physics of spontaneous symmetry breaking in quantum field theory. When the parameter a assumes certain characteristic values (related to the energy eigenvalues of the present system), the solutions, known as Mathieu functions, are periodic with periods of either π or 2π; for arbitrary a the solutions are not periodic (and are not called Mathieu functions). Upon substitution of $z = \cos^2\theta$, the equation assumes the so-called Lindemann form which resembles somewhat the hypergeometric equation [15]. It is quite different, however, having two regular singularities at 0 and 1 and an irregular singularity at ∞. This equation is not one of the factorization types expressible in terms of the generators of the SU(1, 1) dynamical group.

The physical requirement that a quantum mechanical wave function be single-valued—a limitation that does not necessarily apply in a multiply-connected space, as will be seen shortly—and continuous restricts our attention to the periodic solutions or Mathieu functions. These solutions can be classified into four types on the basis of their periodicity and reflection symmetry under Π_x. They are traditionally designated as follows (where $k = 0, 1, 2, \ldots$):

(i) $ce_{2k}(\theta, q)$: even solutions of period π that reduce to $\cos(2k\theta)$ as $q \to 0$; associated eigenvalues are $a = a_{2k}$;

(ii) $ce_{2k+1}(\theta, q)$: even solutions of period 2π that reduce to $\cos[(2k + 1)\theta)]$
as $q \to 0$; associated eigenvalues are $a = a_{2k+1}$;
(iii) $se_{2k+1}(\theta, q)$: odd solutions of period π that reduce to $\sin[(2k + 1)\theta)]$ as
$q \to 0$; associated eigenvalues are $a = b_{2k+1}$; and
(iv) $se_{2k+2}(\theta, q)$: odd solutions of period 2π that reduce to $\sin[(2k + 2)\theta)]$
as $q \to 0$; associated eigenvalues are $a = b_{2k+2}$.

[The symbols ce and se derive from E.T. Whittaker's designations "cosine-elliptic" and "sine-elliptic," respectively.]

Each of the above solutions can be represented by an infinite Fourier series of either $\cos(j\theta)$ or $\sin(j\theta)$—depending on the reflection symmetry—where the integer j is of the same form $(2k, 2k + 1, 2k + 2)$ as that which characterizes the solution. Thus, for example, the even-index Mathieu functions can be represented as

$$ce_{2K}(\theta, q) = \sum_{k=0}^{\infty} A_{2k}^{2K} \cos(2k\theta), \tag{12a}$$

$$se_{2K+2}(\theta, q) = \sum_{k=0}^{\infty} B_{2k+2}^{2K+2} \sin[(2k + 2)\theta]. \tag{12b}$$

The determination of the Fourier coefficients and eigenvalues can be accomplished to arbitrary accuracy through more powerful mathematical techniques than perturbation theory. Two such methods employ continued fractions and the solution of Hill's determinantal equation. There is no general closed-form expression for the eigenvalues, as is the case with the hypergeometric equation and its related forms. This is a consequence of the fact that the Mathieu equation leads to an irreducible three-term, rather than two-term, recursion relation.

With regard to the planar rotator it is important to recognize that the physically significant angular variable is φ, not θ. Thus, the criterion of wave-function continuity, $\psi(\varphi + 2\pi) = \psi(\varphi)$, implies that $\psi(\theta + \pi) = \psi(\theta)$, in which case the admissible solutions are only those Mathieu functions with periodicity π, i.e., $ce_{2k}(\theta, q)$ and $se_{2k+2}(\theta, q)$. From relations (8), (10), and the form of the solutions (12a, b), it is clear that the eigenfunctions of the Hamiltonian and the associated energy eigenvalues can be written as:

$$\psi_{\mu+}(\varphi) = ce_{2\mu}(\varphi/2, \lambda); \qquad E_{\mu+} = \tfrac{1}{4}\varepsilon a_{2\mu} \qquad (\mu = 0, 1, 2, \ldots), \tag{13a}$$

$$\psi_{\mu-}(\varphi) = se_{2\mu}(\varphi/2, \lambda); \qquad E_{\mu-} = \tfrac{1}{4}\varepsilon b_{2\mu} \qquad (\mu = 1, 2, 3, \ldots). \tag{13b}$$

In the limit of vanishing electric field ($\lambda \to 0$), the above solutions reduce to $\cos(\mu\varphi)$ and $\sin(\mu\varphi)$, respectively, i.e., to the coordinate representation of the basis $|\mu\pm\rangle$. This is an important point, for it shows that even though the original basis $|\mu\rangle$, $|-\mu\rangle$ is uncoupled by H_E, the choice of which linear combinations of degenerate states are required for the analytical continuity of a perturbation series is not arbitrary. We shall return to this point again.

From the characteristic values of Mathieu functions (summarized to order

Table I. Characteristic values of Mathieu functions (to order q^6).

$$a_0 = -\frac{1}{2}q^2 + \frac{7}{128}q^4 - \frac{29}{2304}q^6.$$

$$a_1 = 1 + q - \frac{1}{8}q^2 - \frac{1}{64}q^3 - \frac{1}{1536}q^4 + \frac{11}{36\,864}q^5 + \frac{49}{589\,824}q^6.$$

$$b_1 = \text{(substitute } -q \text{ for } q).$$

$$a_2 = 4 + \frac{5}{12}q^2 - \frac{763}{13\,824}q^4 + \frac{1\,002\,401}{79\,626\,240}q^6.$$

$$b_2 = 4 - \frac{1}{12}q^2 + \frac{5}{13\,824}q^4 - \frac{289}{79\,626\,240}q^6.$$

$$a_3 = 9 + \frac{1}{16}q^2 + \frac{1}{64}q^3 + \frac{13}{20\,480}q^4 - \frac{5}{16\,384}q^5 - \frac{1961}{23\,592\,960}q^6.$$

$$b_3 = \text{(substitute } -q \text{ for } q).$$

$$a_4 = 16 + \frac{1}{30}q^2 + \frac{433}{864\,000}q^4 - \frac{5701}{2\,721\,600\,000}q^6.$$

$$b_4 = 16 + \frac{1}{30}q^2 - \frac{317}{864\,000}q^4 + \frac{10\,049}{2\,721\,600\,000}q^6.$$

$$a_5 = 25 + \frac{1}{48}q^2 + \frac{11}{774\,144}q^4 + \frac{1}{147\,456}q^5 + \frac{37}{891\,813\,888}q^6.$$

$$b_5 = \text{(substitute } -q \text{ for } q).$$

$$a_6 = 36 + \frac{1}{70}q^2 + \frac{187}{43\,904\,000}q^4 + \frac{6\,743\,617}{92\,935\,987\,200\,000}q^6.$$

$$b_6 = 36 + \frac{1}{70}q^2 + \frac{187}{43\,904\,000}q^4 - \frac{5\,861\,633}{92\,935\,987\,200\,000}q^6.$$

q^6 in Table I) one can verify the eigenvalues μ_0, $\mu_{1\pm}$ previously obtained by perturbation theory and readily derive the energies of the next two excited states

$$\eta_{2+}^2 = 4^2 + \frac{1}{30}\lambda^2 + \frac{433}{216\,000}\lambda^4 + \cdots, \tag{14a}$$

$$\eta_{2-}^2 = 4^2 + \frac{1}{30}\lambda^2 - \frac{317}{216\,000}\lambda^4 + \cdots. \tag{14b}$$

The unusual broken degeneracy pattern evident in the expressions for the $\mu = 0$, ± 1, ± 2 eigenvalues does in fact extend to all excited levels of the planar rotator: each field-free degenerate pair of states of fixed $|\mu|$ is broken only in order $E^{2\mu}$. Thus, degenerate perturbation theory in sixth order would be needed to break the degeneracy within the $|\mu| = 3$ level. Upon lifting of the degeneracy, it is the states of odd reflection parity that lie lower in energy.

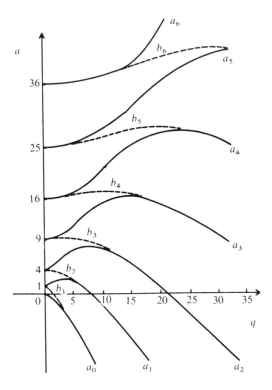

Figure 2.3. Variation of the Mathieu functions a_μ and b_μ as a function of the parameter q (which, for the two-dimensional rotator, depends on the electric or magnetic field strength). (Adapted from Silverman [11].)

From Figure 2.3, which shows the characteristic curves of the Mathieu functions, i.e., the variation of a_μ and b_μ with q, it is seen that the greater the value of $|\mu|$, the greater is the field strength required to effect a noticeable separation between the field-free degenerate states. For sufficiently large values of E, the interaction with the electric field dominates the Hamiltonian, and one would then expect the energy to become negative. This expectation is borne out in Figure 2.3. In the high-q region of the diagram the eigenvalue expansions are no longer valid, and one must utilize the asymptotic properties of the Mathieu functions

$$a_{2k}(q) = b_{2k+1}(q) \sim -2q + (8k + 2)q^{1/2}, \tag{15a}$$

$$a_{2k+1}(q) = b_{2k+2}(q) \sim -2q + (8k + 6)q^{1/2},$$

$$(k = 0, 1, 2, \ldots; q \text{ large and positive}), \tag{15b}$$

to determine the energy spectrum. The resulting energy eigenvalues take the form

$$E_{\mu\pm} = -|e|RE + (\mu \pm \tfrac{1}{2})\hbar(|e|E/mR)^{1/2}, \tag{16a}$$

which is amenable to a simple physical interpretation. The first term, which is negative and dominates the expression for large E, represents the energy of a classical electric dipole $\mu_E = |e|R$ aligned parallel to the field. The second term characterizes a harmonic oscillator spectrum with equidistant level separations of $\hbar\omega$ where

$$\omega = (|e|E/mR)^{1/2} = (\mu_E E/I)^{1/2} \tag{16b}$$

is the oscillation angular frequency of a dipole μ_E with moment of inertia $I = mR^2$. The reason for this behavior will become apparent when we examine the eigenfunctions.

Before concluding this discussion of the energy spectrum of the planar rotator in an electric field. I would like to clarify an important general point concerning degenerate perturbation theory and the symmetry properties of the Hamiltonian. In particular, consider the seemingly paradoxical question: Why does one use degenerate perturbation theory when treating the interactions of degenerate states? It would seem that the answer is obvious. Nondegenerate perturbation theory fails, as is signaled by the appearance of terms with vanishing energy denominators. Degenerate perturbation theory eliminates these terms through construction of linear combinations of unperturbed states for which the matrix elements in the numerator likewise vanish. Thus, singular terms do not appear.

Suppose, however, that the degenerate basis with which one begins is already uncoupled by the interaction; this is the case with the planar rotator since $\langle -\mu|H_E|\mu\rangle = 0$. Under these circumstances, where singularities do not occur, it may seem—and the pedagogical literature has encouraged this view [5]—that it is permissible to use nondegenerate perturbation theory. This assertion cannot be correct, for the application of nondegenerate perturbation theory to the two-dimensional rotator leads to energy eigenvalues that are functions of μ^2 and therefore remain doubly degenerate to all orders. The error lies in the failure to recognize the primary objective of degenerate perturbation theory which is not merely to eliminate singular terms, but to select unambiguously the appropriate linear combination of degenerate states from among an infinite number of possibilities. To start with any other combination leads to a discontinuous change of states in the limit of vanishing interaction parameter, so that the perturbative expansions are not valid.

But how is one to know whether or not a particular interaction term—let us call it H'—removes the original degeneracy in some order? This knowledge is presumably the outcome of the calculation—yet which calculation one performs (degenerate versus nondegenerate perturbation theory) seems to depend on it. For the case of the planar rotator the fact that the eigenvalues of degenerate perturbation theory agree with the characteristic values of the Mathieu functions is proof that the degeneracy is indeed lifted and that one must construct an appropriate initial basis ($\cos \mu\varphi$, $\sin \mu\varphi$) even though the matrix elements between the original basis ($e^{i\mu\varphi}$, $e^{-i\mu\varphi}$) vanish. What must one do, however, when an exact solution is not known? Would it actually have been necessary to employ eighth-order perturbation theory to discover that the degeneracy is lifted between $|\mu| = 4$ states?

It is at this point that the theory of groups comes to the rescue. One must examine the symmetries of the unperturbed Hamiltonian H_0 and the perturbation H' in order to ascertain whether the original degenerate basis, which spans an irreducible representation of the symmetry group H_0, remains an irreducible representation of the (possibly lower) symmetry group of $H_0 + H'$.

In the case of the planar rotator the Hamiltonian $H_0 = L_z^2/2mR^2$ is invariant under the elements of the continuous group SO(2), the group of rotations in the plane. It is also invariant under the discrete symmetry group comprising the identity $\mathscr{I}$, the reflections Π_x and Π_y, and the inversion Π; this group is isomorphic to the four-group (*Vierergruppe*) V. The total symmetry of H_0 is the two-dimensional rotation–reflection group, a mixed continuous group that can be parametrized by a continuous parameter φ (the rotation angle) and a discrete parameter $d = \pm 1$ (a determinant). One matrix representation of this group is

$$\{\varphi, d\} = \begin{pmatrix} \cos \varphi & \sin \varphi \\ -d \sin \varphi & d \cos \varphi \end{pmatrix}, \tag{17a}$$

where

$$\{\varphi, d\} \cdot \{\varphi', d'\} = \{d'\varphi + \varphi', dd'\}, \tag{17b}$$

gives the group composition function [16]. This group is clearly non-abelian (i.e., the group operations do not commute) although the subgroups SO(2) and V are abelian. The basis $e^{\pm i\mu\varphi}$ spans a two-dimensional representation for $\mu \neq 0$.

Upon addition of an electric field **E** perpendicular to the rotation axis the total Hamiltonian $H = H_0 + H'$ (with $H' = -\boldsymbol{\mu_E} \cdot \mathbf{E}$) is invariant under a much smaller symmetry group comprising only the elements $\mathscr{I}$ and Π_x. This group is isomorphic to the so-called symmetric or permutation group S_2 which is abelian and therefore can have only one-dimensional representations. There are two such representations designated, respectively, Γ^S (which is symmetric under Π_x) and Γ^A (which is antisymmetric under Π_x). A simple character analysis [17] shows that if $\Gamma_{\mu d}$ is an irreducible representation of the rotation–reflection group, then $\Gamma_{\mu d} \to \Gamma^S + \Gamma^A$ upon the lowering of the symmetry to S_2 by addition of the electric field. Thus, the degenerate basis splits under the action of **E**; the exact solutions $ce_{2\mu}(\varphi/2, \lambda)$ and $se_{2\mu}(\varphi/2, \lambda)$ display the group-theoretically predicted symmetries.

Group-theoretical arguments cannot in general give the scale of an interaction, i.e., they cannot yield the magnitude of the splitting. The power of group theory is that it can give the degeneracy pattern which, moreover, is valid to all orders of perturbation theory. An important example is the three-dimensional rigid rotator in an electric field along the rotation axis [18]. Group theory shows that the degeneracy of states with different angular momentum quantum numbers l and $|\mu|$ is lifted, but, for given l, there is a residual degeneracy of states differing only in the sign of μ; this degeneracy is never lifted as long as the axial symmetry is not broken.

Mention should also be made of so-called accidental degeneracies [19], i.e., the occurrences of degenerate states belonging to different irreducible representations of the symmetry group of the Hamiltonian. The hydrogen atom, or more precisely the spinless Coulomb problem, is a well-known example; the degeneracy among the n^2 states of fixed principal quantum number n and different orbital quantum numbers l cannot be explained on the basis of the most obvious symmetry group, SO(3), the three-dimensional rotation group. In such cases one can usually find a larger invariance group that does account for the degeneracies. For hydrogen this group is SO(4), the four-dimensional rotation group. Accidental degeneracy, while characteristic of a specific dynamical law, is not necessarily an unexplainable degeneracy.

In short, then, if by group-theoretical or other means it is clear that an interaction will lift the degeneracy of an initially degenerate set of states, nondegenerate perturbation theory cannot be used even though the states of the basis may not be directly coupled by the interaction.

Let us consider next the wave functions and corresponding electron distributions of the planar rotator. The exact solutions (to within a normalization constant) are the Mathieu functions given by (13a, b) which are representable in the following Fourier decompositions to order q^2:

$$ce_k(\varphi, q) = \cos k\varphi - \tfrac{1}{4}q\left(\frac{\cos(k+2)\varphi}{k+1} - \frac{\cos(k-2)\varphi}{k-1}\right)$$

$$+ \tfrac{1}{32}q^2\left(\frac{\cos(k+4)\varphi}{(k+1)(k+2)} + \frac{\cos(k-4)\varphi}{(k-1)(k-2)}\right), \qquad (18a)$$

$$se_k(\varphi, q) = \sin k\varphi - \tfrac{1}{4}q\left(\frac{\sin(k+2)\varphi}{k+1} - \frac{\sin(k-2)\varphi}{k-1}\right)$$

$$+ \tfrac{1}{32}q^2\left(\frac{\sin(k+4)\varphi}{(k+1)(k+2)} + \frac{\sin(k-4)\varphi}{(k-1)(k-2)}\right), \qquad (18b)$$

valid for $q^2/[2(k^2 - 1)] \ll k^2$ with $k > 0$. To first order in q (or λ) these expansions correspond to the results of first-order perturbation theory for which, as emphasized previously, analytical continuity at $\lambda = 0$ requires the use of the appropriate zero-field states $|\mu\pm\rangle$.

The ground-state wave function, obtained from first-order perturbation theory, is

$$\psi_{0+}(\varphi) = (1 - \lambda \cos \varphi)/\sqrt{2\pi}, \tag{19a}$$

and leads to the probability distribution (to order λ)

$$P_{0+}(\varphi) = |\psi_{0+}(\varphi)|^2 = (1 - 2\lambda \cos \varphi)/2\pi. \tag{19b}$$

Thus, the electron density is seen to be greatest at $\varphi = \pi$ and least at $\varphi = 0$. This can be understood on the basis of a classical model; a static electric dipole will align itself along the field so as to minimize its energy. The wave functions of the excited states, deducible from relations (18a, b), lead again in first order to the probability densities

$$P_{\mu+}(\varphi) = \frac{1}{\pi}\left[\cos^2(\mu\varphi)\left(1 + \frac{2\lambda}{4\mu^2 - 1}\cos\varphi\right) + \frac{2\mu\lambda}{4\mu^2 - 1}\sin(2\mu\varphi)\sin\varphi\right], \tag{19c}$$

$$P_{\mu-}(\varphi) = \frac{1}{\pi}\left[\sin^2(\mu\varphi)\left(1 + \frac{2\lambda}{4\mu^2 - 1}\cos\varphi\right) - \frac{2\mu\lambda}{4\mu^2 - 1}\sin(2\mu\varphi)\sin\varphi\right]. \tag{19d}$$

Recall that in perturbation theory wave functions to order λ^{k-1} determine the energy eigenvalues to order λ^k. Since degeneracy breaking in the present case does not occur until order $\lambda^{2\mu}$ for degenerate states $|\pm\mu\rangle$, the wave functions characterizing nondegenerate states most be of order $\lambda^{2\mu-1}$. To order λ^1, therefore, (19c, d) represent a pair of degenerate states.

The physical interpretation of the above excited-state probability densities which characterize a rotating electric dipole is not as straightforward as the interpretation of the ground-state distribution characterizing a stationary dipole. For small λ the dominating feature of the probability densities is the $\cos^2(\mu\varphi)$ and $\sin^2(\mu\varphi)$ dependence in the states of even and odd reflection parity, respectively. The average probability distribution of the two states, however,

$$\bar{P}_\mu = (P_{\mu+} + P_{\mu-})/2 = \frac{1}{2\pi}\left[1 + \frac{2\lambda}{4\mu^2 - 1}\cos\varphi\right] \tag{19e}$$

yields a simpler expression that is amenable to interpretation according to a classical model. Equation (19e) predicts a maximum electron density at $\varphi = 0$ and a minimum density at $\varphi = \pi$ in direct opposition to the ground-state distribution. One can account for this by arguing that the classical rotating dipole passes the minimum potential region around $\varphi = \pi$ with a higher speed than that with which it passes the maximum potential region around $\varphi = 0$. Hence the dipole spends less time in the vicinity of $\varphi = \pi$ than it does in the region of $\varphi = 0$. It is of interest to note that by incorrectly applying nondegenerate perturbation theory to the planar rotator one obtains (19e) directly as the excited-state electron density.

For high-field strengths the perturbation expansions in λ are no longer valid, and one must utilize asymptotic expressions for the Mathieu functions which, for sufficiently large and positive q, can be written in terms of elliptic functions or Hermite polynomials. These expressions are not particularly

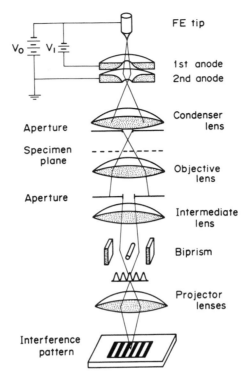

Figure 1.1. Electron interference with a field-emission electron microscope. Wave front splitting occurs at the biprism. (Courtesy of A. Tonomura, Hitachi ARL.)

(effectively the two slits of the apparatus), the two components of the electron beam recombined in the observation plane of the microscope where the build-up of a pattern of interference fringes was recorded on film and on a TV monitor.

The appearance of a fringe pattern is not in itself extraordinary. After all, if electrons were actually waves, then the experimental configuration would represent a type of wavefront-splitting interferometer, and there is nothing unusual about the linear superposition of waves to generate an interference pattern. What is startling, however, is the observed *emergence* of the fringe pattern in a microscope of approximately 1.5 m length under conditions where the mean interval between successive electrons is over 100 km! Clearly a given electron had been detected long before the succeeding electron was "born" at the field-emission tip. Under these circumstances it is unlikely that there can be any sort of cooperative interaction between the electrons of the beam.

Electron detection events appear on the TV monitor one by one at random locations as illustrated in Figure 1.2. The first few hundred scattered spots hardly hint at any organization. However, by the time some hundred thousand electrons have been recorded, stark alternating stripes of white and black stand out sharply as if made by the two-slit interference of laser light.

illuminating and will not be discussed here further. The most important attribute of the asymptotic Mathieu functions, however, is that they are significant only in the vicinity of $\theta = \pm \pi/2$, dropping off rapidly to very small values near $\theta = 0, \pi$ as q increases. Thus the strong-field wave functions of both reflection symmetries become more tightly confined about $\varphi = \pi$ as the field strength increases. The system is again interpretable in terms of a classical dipole aligned closely along the field direction. The dipole is not stationary, however, but undergoes small-amplitude oscillations about the equilibrium position. This is readily confirmed by expanding the Hamiltonian, (3b), about the point $\varphi = \pi$ to obtain the Hamiltonian of a harmonic oscillator. In this way the occurrence of an oscillator spectrum in (16a) is accounted for.

2.3. The Two-Dimensional Rotator in a Magnetic Field

The interaction of the planar rotator with a magnetic field (Zeeman effect) provides some instructive similarities and contrasts to the previously discussed behavior in an electric field. As before, symmetry considerations can reveal significant features of the system before any detailed analytic solution of the equation of motion is attempted. Let the rotator (of radius R) again be in the x–y plane with the magnetic field **B** in the y–z plane at angle β to the rotation axis z as shown in Figure 2.1. The position vector of the electron is completely determined by the rotation angle φ with respect to the x axis.

The total Hamiltonian of the system is then expressible in the form

$$H = H_0 + H_M = L_z^2/2mR^2 - \omega_L L_z \cos \beta + \tfrac{1}{2} m\omega_L^2 R^2(\cos^2 \varphi + \sin^2 \varphi \cos^2 \beta),$$
$$(20a)$$

where

$$\omega_L = -eB/2mc \qquad (20b)$$

is the Larmor frequency (a positive number for a negative charge). The term in H_M linear in ω_L and independent of the electron coordinate is the paramagnetic term; the diamagnetic contribution to H_M is quadratic in ω_L and proportional to the component of the magnetic field normal to the electron position vector. As seen from (20a), there are two inequivalent basic orientations of the magnetic field that can influence the state of motion of the electron within the plane of rotation:

 (i) **B** along the rotation axis; and
 (ii) **B** normal to the rotation axis.

In the case of the Stark effect there is only one orientation of the electric field; an electric field along the rotation axis merely results in a constant potential that can be set equal to zero by locating the electron in the plane $z = 0$.

The case of **B** parallel to the rotation axis, however, is trivially solvable, for the energy eigenvectors are simply the eigenstates $|\mu\rangle$ of L_z. The eigenvalues are then

$$E_\mu = \varepsilon\mu^2 - \mu\hbar\omega_L + m\omega_L^2 R^2/2 \qquad (\mu = 0, \pm 1, \pm 2, \dots), \qquad (21)$$

where ε has been defined in (3c). Thus, an axially oriented magnetic field lifts the double degeneracy of the excited states, whereas an axial electric field has no effect. It is perhaps not superfluous to point out the reason for this. An electric field (a polar vector) represents a preferential *direction*. By contrast, a magnetic field (an axial vector) is a chiral structure with a preferential *sense* (clockwise or counterclockwise) or *handedness* (left or right). These characteristics, of which much more will be said in Chapter 6, stem from the origin of the two fields: an electric field is produced by charges, whereas a magnetic field arises from currents. The states $|\pm|\mu|\rangle$ have probability current densities that circulate parallel and antiparallel, respectively, to the current producing the magnetic field and therefore represent two inequivalent orientations of a magnetic dipole in the magnetic field.

The case of **B** lying within the plane of rotation ($\beta = \pi/2$) is not trivially solvable and will be the focus of our attention in this section. In the analysis of the rotator in an electric field, it was shown that the Stark solutions span the symmetric and antisymmetric one-dimensional irreducible representations of a group isomorphic to the symmetric group S_2. An examination of the Hamiltonian including the rotational kinetic energy and diamagnetic potential energy—there is no paramagnetic term (term linear in the Larmor frequency) for the chosen field orientation—shows that it is invariant under a group of operations comprising the identity, reflection across an axis in the rotation plane normal to the field, and inversion through the origin. This group is isomorphic to the (abelian) four-group V. Since the degenerate field-free basis spans a two-dimensional irreducible representation of the rotation-reflection group, it must split under the magnetic interaction into nondegenerate states that span the irreducible representations of V. A simple character analysis shows that states of even or odd $|\mu|$, respectively, span representations of even or odd inversion parity. However, the two states of the same inversion parity that belong to a rotational level of given $|\mu|$ differ in their reflection parities. One would therefore expect the Zeeman effect to give rise to stationary states of four possible symmetry types in contrast to the two possible symmetry types of the Stark effect. The higher symmetry also leads to marked differences in the energy spectra between the two cases, particularly in the high-field domain. Let us now examine this system in more detail.

In a coordinate representation the Hamiltonian (20a) without paramagnetic term can be written in the form

$$H = -\varepsilon(\partial^2/\partial\varphi^2 - \xi^2\cos^2\varphi), \qquad (22a)$$

where the magnetic interaction parameter ξ is

$$\xi = m\omega_L R^2/\hbar. \tag{22b}$$

With energy eigenvalues again expressed in the form of relation (8) (but with the appropriate quantum labels to be specified), Hamiltonian (22a) leads to a Schrödinger equation which can be cast once more into the canonical form of Mathieu's equation

$$d^2\psi_\eta/d\varphi^2 + (a - 2q\cos 2\varphi)\psi_\eta(\varphi) = 0, \tag{23a}$$

where

$$a = \mu^2 - \xi^2/2, \tag{23b}$$

and

$$q = \xi^2/4. \tag{23c}$$

Note that the physically significant angular coordinate φ appears, and not $\theta = \varphi/2$ as in the case of the Stark effect, (11). As a consequence, the boundary condition $\psi(\varphi + 2\pi) = \psi(\varphi)$ imposed by the continuity of the wave function can be satisfied by Mathieu functions of both 2π and π periodicity.

The exact solutions (up to a normalization constant) are therefore the four types of Mathieu functions described earlier, and the wave functions and energy eigenvalues can be written as

$$\psi_{\mu+}^\pi(\varphi) = ce_\mu(\varphi, \xi); \qquad E_{\mu+}^\pi = \varepsilon(a_\mu(\xi) + \xi^2/2) \qquad (\mu = 0, 2, 4, \ldots), \tag{24a}$$

$$\psi_{\mu-}^\pi(\varphi) = se_\mu(\varphi, \xi); \qquad E_{\mu-}^\pi = \varepsilon(b_\mu(\xi) + \xi^2/2) \qquad (\mu = 2, 4, 6, \ldots), \tag{24b}$$

$$\psi_{\mu+}^{2\pi}(\varphi) = ce_\mu(\varphi, \xi); \qquad E_{\mu+}^{2\pi} = \varepsilon(a_\mu(\xi) + \xi^2/2) \qquad (\mu = 1, 3, 5, \ldots), \tag{24c}$$

$$\psi_{\mu-}^{2\pi}(\varphi) = se_\mu(\varphi, \xi); \qquad E_{\mu-}^{2\pi} = \varepsilon(b_\mu(\xi) + \xi^2/2) \qquad (\mu = 1, 3, 5, \ldots), \tag{24d}$$

where, in addition to eigenvalues of L_z, the states are labeled by their periodicity and reflection symmetry with respect to Π_x.

For a weak magnetic field the eigenvalues can be expanded in a perturbation series in the parameter ξ^2. From Table I, the energy of the ground state and first six excited states, truncated at the lowest order in ξ^2 that breaks the degeneracy, is as follows:

$$(\eta_{0+}^\pi)^2 = \tfrac{1}{2}\xi^2, \tag{25a}$$

$$(\eta_{1+}^{2\pi})^2 = 1 + \tfrac{3}{4}\xi^2, \tag{25b}$$

$$(\eta_{1-}^{2\pi})^2 = 1 + \tfrac{1}{4}\xi^2, \tag{25c}$$

$$(\eta_{2+}^\pi)^2 = 4 + \tfrac{1}{2}\xi^2 + \tfrac{5}{192}\xi^4, \tag{25d}$$

$$(\eta_{2-}^\pi)^2 = 4 + \tfrac{1}{2}\xi^2 - \tfrac{1}{192}\xi^4, \tag{25e}$$

$$(\eta_{3+}^{2\pi})^2 = 9 + \tfrac{1}{2}\xi^2 + \tfrac{1}{256}\xi^4 + \tfrac{1}{4096}\xi^6, \tag{25f}$$

$$(\eta_{3-}^{2\pi})^2 = 9 + \tfrac{1}{2}\xi^2 + \tfrac{1}{256}\xi^4 - \tfrac{1}{4096}\xi^6. \tag{25g}$$

From the above relations one sees that the double degeneracy of zero-field states of specified $|\mu|$ is broken by a magnetic field (in the rotator plane) in order ξ^μ in contrast to the Stark effect in which degeneracy breaking occurs in order $\lambda^{2\mu}$.

When the magnetic field is sufficiently great that the eigenvalue expansions are no longer valid, one must again turn to the asymptotic properties of the Mathieu functions. From (15a, b) and (23b, c) one finds that the quadratic dependence of the energy on ξ (or B) drops out, and the eigenvalue spectrum again assumes a form highly suggestive of that of a harmonic oscillator:

$$E_{\mu+}^{2\pi} = (\mu + \tfrac{1}{2})\hbar\omega_L \qquad (\mu = 0, 2, 4, \ldots), \tag{26a}$$

$$E_{\mu-}^{2\pi} = (\mu - \tfrac{1}{2})\hbar\omega_L \qquad (\mu = 2, 4, 6, \ldots), \tag{26b}$$

$$E_{\mu\pm}^{\pi} = (\mu \pm \tfrac{1}{2})\hbar\omega_L \qquad (\mu = 1, 3, 5, \ldots), \tag{26c}$$

An interesting and significant distinction between the high-field Zeeman and Stark spectra is that states split by an electric field E remain nondegenerate for all values of E, but a very strong magnetic field actually recreates a double degeneracy as evident in Figure 2.3, and analytically verified in relations (26a–c). A given degenerate level consists of two states of opposite reflection symmetry and periodicity. In the next higher level the periodicity labels are reversed. Thus, for example, the lowest level consists of the states ce_0^π and $se_1^{2\pi}$; the first excited level consists of the states $ce_1^{2\pi}$ and se_2^π. Every two successive levels exhaust all four types of Mathieu functions.

These high-field states give rise to a quasi-Landau energy spectrum. The true Landau energies for a free electron orbiting in a plane perpendicular to a static uniform magnetic field are

$$E_n = (n + \tfrac{1}{2})\hbar\omega_C \qquad (n = 0, 1, 2, \ldots), \tag{27a}$$

where

$$\omega_C = 2\omega_L \tag{27b}$$

is the cyclotron frequency. Differences between the spectra of (26a–c) and (27a, b) are to be expected since the two physical systems are not exactly comparable, differing in both the orientation of the magnetic field (normal versus parallel to the rotation axis) and the accessible space (a circular perimeter versus the entire x–y plane). The resemblance of the Landau or quasi-Landau spectrum to the harmonic oscillator spectrum is again not merely fortuitous, but a consequence of the fact that the diamagnetic contribution to the Hamiltonian (20a) is of the form of a harmonic oscillator potential.

Although the exact solutions given above allow one to deduce all the properties of the planar rotator in a magnetic field, several important physical concepts can be further clarified by examining the system from the perspectives of both perturbation theory and group theory, as was done with the Stark effect. In the absence of a paramagnetic term, the matrix elements of H_M in the field-free basis $|\mu\rangle$ are

$$\langle\mu|H_M|\mu'\rangle = \tfrac{1}{2}\varepsilon\xi^2[\delta_{\mu,\,\mu'} + \tfrac{1}{2}(\delta_{\mu,\,\mu'+2} + \delta_{\mu,\,\mu'-2})], \tag{28}$$

from which it is clear that the diagonal elements of H_M are nonvanishing in contrast to the diagonal elements of H_E. This indicates that the states $|\mu\rangle$ already constitute a field-free basis with nonvanishing magnetic dipole moment. A comparable electric dipole moment is precluded by the symmetry of the Hamiltonian (3a). The off-diagonal elements $(H_M)_{-\mu,\,\mu}$ are in general zero. There is an exception, however, for $\mu = \pm 1$ where, for each choice of sign, one term in (28) survives. As a consequence of this coupling, the double degeneracy of the first excited level is split by first-order degenerate perturbation theory in contrast to the Stark effect. A splitting linear in H_M, however, is still quadratic in the magnetic field. Thus, the symmetry of the planar rotator is such that, despite the degeneracy of its excited states, no splitting linear in either the electric or magnetic field occurs for a weak field normal to the rotation axis. The degeneracy of the states with $|\mu| > 1$ can be removed by successively higher orders of perturbation theory, and these calculations must yield, of course, the same expansions as obtained from the characteristic values of the Mathieu functions.

Having begun the discussion of the magnetic interaction of the planar rotator with a brief remark on symmetry, let us conclude by noting once more the power of group theory to provide a more fundamental way of understanding the correlations of state symmetries with rotational excitation other than by solving the Schrödinger equation and discovering these correlations in the Mathieu function solutions. In particular, I address the questions why is it possible for a strong magnetic field to recreate degeneracy among states, whereas the degeneracy lifted by an electric field remains broken independent of the field strength, and what leads to the correlations in periodicity and reflection symmetry of the strong-field states that become degenerate? The answer lies in the symmetry of the Hamiltonian.

The rotational kinetic energy H_0, as discussed before, is invariant under the two-dimensional rotation-reflection group, and the zero-field basis $e^{\pm i\mu\varphi}$ spans a two-dimensional irreducible representation $\Gamma_{\mu d}$ of this group. The total Hamiltonian (20a), as pointed out earlier, is invariant under a smaller group of symmetry operations, the abelian four-group or V, comprising the elements $\mathscr{I}$, Π_x, Π_y, and Π. It is clear, therefore, that the degenerate basis must split in a magnetic field although, since V is a higher symmetry group than S_2 and has twice as many irreducible representations, it is perhaps not so obvious which zero-field states go into which irreducible representations. Using the character table of the four-group and the orthogonality properties of group characters, one can verify that the degeneracy breaking must occur as follows:

$$\Gamma_{\mu d} \to \begin{cases} \Gamma^1 + \Gamma^2, & \mu \text{ even} \\ \Gamma^3 + \Gamma^4, & \mu \text{ odd} \end{cases}, \tag{29}$$

Table II. Character table of the four group.

V	$\mathscr{I}$	Π_x	Π_y	Π	Basis $(n = 0, 1, 2, \ldots)$
Γ^1	1	1	1	1	$ce_{2n}^{(\pi)}$
Γ^2	1	-1	-1	1	$se_{2n+2}^{(\pi)}$
Γ^3	1	1	-1	-1	$ce_{2n+1}^{(2\pi)}$
Γ^4	1	-1	1	-1	$se_{2n+1}^{(2\pi)}$
Γ_{nd}	2	0	0	$(-2)^n$	$e^{in\varphi}, e^{-in\varphi}$

where the irreducible representations are defined and correlated with the Mathieu functions in Table II. The above decomposition implies that states of even or odd rotation quantum number μ are, respectively, split into states of even or odd inversion parity. Since $\Pi_x \Pi_y = \Pi$, even (odd) inversion parity comes about only if the reflection parities of the states are, respectively, the same (opposite). Translated into the symmetry properties of the Mathieu functions, the above group theoretical results require that field-free states of even or odd μ split into pairs ce_μ, se_μ with, respectively, even or odd μ. These split pairs remain split independently of the strength of the magnetic field.

However, there is *no* group-theoretic restriction against a basis state of Γ^1 or Γ^2 becoming degenerate with a basis state of Γ^3 or Γ^4 as the field strength increases. This follows from the fact that these two pairs of representations arise from different irreducible representations of the rotation–reflection group. Thus, *should degeneracy occur*, group theory requires that the degenerate states occur in the pairs $(ce_{2\mu}, se_{2\mu+1})$ and $(ce_{2\mu+1}, se_{2\mu+2})$ where, for both cases, $\mu = 0, 1, 2$, etc.

Group theory cannot say whether a degeneracy in this case will occur—and, in fact, an *exact* degeneracy in the high-field spectrum does not really occur. The asymptotic relation $a_\mu \approx b_{\mu+1}$ between the characteristic values of the Mathieu functions holds ideally only in the limit $q \to \infty$; for finite q the difference $b_{\mu+1} - a_\mu$ falls off exponentially with the square root of q, a difference that is effectively negligible for the high values of q at which the asymptotic formulas are valid.

We will not consider the properties of the Zeeman wave functions in the same detail as for the Stark effect, since the discussions would be similar. From the Fourier decompositions given by relations (18a, b) one can construct the perturbation series for each state. The basic distinction to keep in mind is that the angular variable which appears is now φ, the electron coordinate. This has an important consequence for the high-field asymptotic wave functions. As discussed in the previous section, the Mathieu functions for large q assumed their maximum values in the vicinity of $\theta = \pm \pi/2$ and dropped rapidly to small values at $\theta = 0, \pi$ which represented the same location. Since $\theta = \varphi$ in the present case, the electron density is greatest about the *two* points where the magnetic field axis intersects the orbital circle. Expansion of the Hamiltonian about these points again results in the Hamiltonian of a harmonic oscillator. Interpreted classically, the rotator can no longer rotate freely, but executes small-amplitude oscillations about the direction either parallel or antiparallel to the field. This is consistent with the observation that the Mathieu functions are linear superpositions of components with both positive and negative magnetic dipole moments.

2.4. The Two-Dimensional Rotator in a Vector Potential Field

We have seen that the presence of electric and magnetic fields alters the energy and wave functions of a charged particle confined to a circular orbit (or, more precisely, to a linear space of constant curvature $1/R$). They break the level degeneracy in interesting ways that can be elucidated by symmetry arguments and give rise to a range of motions that, depending on field strength, vary between free rotation with well-defined angular momentum and harmonic oscillation about points of equilibrium. While there are aspects of the interaction of the two-dimensional rotator with electromagnetic fields that can be correctly determined only by quantum mechanics, the system nevertheless has classical counterparts in the behavior of electric and magnetic dipoles which, at least in the domain of applicability of the Bohr correspondence principle, yield quantitative results comparable to those arrived at quantum mechanically.

Consider next, however, the configuration of Figure 2.4 for which—depending on the history of the system—there may or may not be a classical analogue. This is the bound-state counterpart of the AB effect treated earlier in the context of a charged particle beam. The electron is contrained to a circular space of radius R through which is now threaded, so to speak, a long tube of magnetic flux. We can imagine the flux confined to the interior of a long solenoid of radius $a \ll R$. Then, as in Chapter 1, the circulating electron is in an environment permeated by a vector potential field (specified in the Coulomb gauge by (1.9)) ideally free of electric and magnetic fields, excluding the fields of the particle itself. Although there is no "splitting" of wave packets in this case, and hence no recombination and interference to speak of, there can still be a physical effect of the local vector potential—or remote magnetic field—on the state of motion of the orbiting particle. The nature of this effect is the issue we will examine.

Since there is a certain measure of arbitrariness in the specification of a gauge field like the magnetic vector potential, it is pertinent to comment briefly here on the matter of gauge invariance and observability of a dynamical variable. It is well known that to be an observable, a dynamical variable must be representable by a Hermitian operator with a complete set of eigenstates [20]. For a mechanical system coupled to a gauge field, a

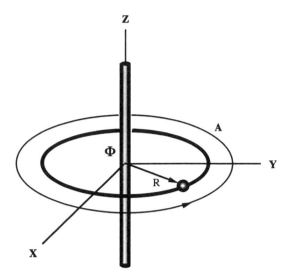

Figure 2.4. Configuration of the two-dimensional charged rotator in a vector potential field manifesting a bound-state AB effect.

further criterion must be met so that physically meaningful quantities do not depend on the arbitrary choice of gauge. All theoretical expressions, e.g., quantum mechanical expectation values and transition matrix elements, representing measurable quantities must be invariant under a gauge transformation as summarized by relations (1.5a, b, c). In the case of a charged particle orbiting a tube of magnetic flux, a gauge transformation cannot change the magnetic field inside the tube if the vector potentials outside the tube satisfy Stokes's theorem

$$\oint_C \mathbf{A} \cdot \mathbf{ds} = \oint_C \mathbf{A'} \cdot \mathbf{ds} = \text{total magnetic flux} = \Phi, \tag{30a}$$

where the contour circumscribes the solenoid. It then follows from (1.5a), that

$$\oint_C \nabla\Lambda \cdot \mathbf{ds} = \oint_C d\Lambda = 0, \tag{30b}$$

and the gauge function $\Lambda(\mathbf{r}, t)$ is single-valued, a mathematical property of importance in previous defenses of the theoretical existence of the AB effect [21].

Under a gauge transformation the (nonrelativistic) Hamiltonian

$$H = (\mathbf{p} - e\mathbf{A}/c)^2/2m + e\phi, \tag{31a}$$

and Schrödinger equation

$$H\Psi = i\hbar\,\partial\Psi/\partial t, \tag{31b}$$

are form-invariant, the transformed Hamiltonian being given by relation
(1.7c) with $U = e^{ie\Lambda/\hbar c}$ as follows:

$$H' = e^{ie\Lambda/\hbar c}He^{-ie\Lambda/\hbar c} - (e/\hbar c)\,\partial\Lambda/\partial t = (\mathbf{p} - e\mathbf{A}'/c)^2/2m + e\phi'. \quad (31c)$$

Thus, the time-evolution of the physical system is unaffected by a gauge
transformation, i.e., Ψ' evolves under H' in the same way that Ψ evolves
under H.

The eigenvalue spectrum of H, however, is not invariant under a gauge
transformation. If $\Psi_E = \Psi_E^0 \exp(-iEt/\hbar)$ is a stationary state of H with
eingenvalue E, then $\Psi'_E = \exp(ie\Lambda/\hbar c)\Psi_E$ is also an eigenstate, but not
necessarily a *stationary* state, of H' with eigenvalue $E' = E - (e/\hbar c)\,\partial\Lambda/\partial t$.
Thus, E' is both gauge- and time-dependent unless Λ is independent of time.
If the latter condition holds, the Hamiltonians H' and H are simply related
by a unitary transformation and necessarily have the same eigenvalue
spectrum.

Since the eigenvalues of H are not gauge-invariant, does this imply that
energy is not an observable? The answer is clearly "no," for the reason that
in quantum mechanics, as in classical mechanics, the Hamiltonian need not
represent the energy of the system (a known, but not widely appreciated,
point). A suitable operator that *does* represent the energy of the system
should have a time-independent scalar potential $\phi(x)$. Under an arbitrary
gauge transformation, the true energy operator [22], W, whose eigenvalues
are obtained from

$$W\Psi = E\Psi, \quad (32a)$$

and characterize the allowed energies of the system, becomes

$$W = H' + (e/\hbar c)\,\partial\Lambda/\partial t = e^{ie\Lambda/\hbar c}He^{-ie\Lambda/\hbar c}. \quad (32b)$$

It is evident that the eigenvalue spectrum and expectation values of W are
independent of gauge and correspond to observable quantities. W, of course,
does not ordinarily generate the time evolution of the system; that is the role
of the Hamiltonian.

The Hamiltonian (31a) of the planar rotator in the vector potential
field (1.9) of a tube of magnetic flux (with $\phi = 0$) can be succinctly
written in the form

$$H = \varepsilon K_z^2 = \varepsilon(L_z - \alpha)^2, \quad (33a)$$

where K_z is the component of the kinetic angular momentum

$$\mathbf{K} = \mathbf{r} \times (\mathbf{p} - e\mathbf{A}/c) \quad (33b)$$

normal to the rotator plane (the only component) and

$$\alpha = \Phi/\Phi_0 \quad (33c)$$

is the ratio of enclosed magnetic flux to the fluxon Φ_0 (see (1.10c)).

Since the distinctions between canonical and kinetic dynamical variables have at times been the source of some confusion, it is worth noting here the properties and roles of **K** and the canonical orbital angular momentum **L**. The components of **L**, which satisfy the commutation relations

$$[L_i, L_j] = i\hbar e_{ijk}L_k, \tag{34a}$$

where e_{ijk} is the Levi-Civita permutation symbol (or the completely anti-symmetric tensor), are the generators of an infinitesimal rotational transformation of angle θ about direction **n** according to

$$\langle \mathbf{r}|(1 + i\mathbf{n}\cdot\mathbf{L}\,d\theta/\hbar) = \langle \mathbf{r} + (\mathbf{n} \times \mathbf{r})\,d\theta|. \tag{34b}$$

The matrix elements of **L**, however, are not invariant under a gauge transformation, but transform as follows:

$$\langle \mathbf{L}\rangle_{\Psi'} = \langle \mathbf{L}\rangle_{\Psi} + (e/\hbar c)\langle \mathbf{r} \times \nabla\Lambda\rangle_{\Psi}, \tag{34c}$$

where the bracket represents either a transition moment or expectation value. **L**, therefore, ceases in general to be an observable in a mechanical system coupled to an electromagnetic vector potential. By contrast, the dynamical variable **K** represents the mechanical angular momentum of a particle (corresponding, for example, to mvR for a classical particle in uniform circular motion); it has gauge-invariant matrix elements and satisfies a different set of commutation relations [23]

$$[K_i, K_j] = i\hbar e_{ijk}[K_k + (e/\hbar c)x_k(\mathbf{x}\cdot\nabla \times \mathbf{A})]. \tag{35}$$

For a particle in the field-free region outside the solenoid (where $\mathbf{B} = \nabla \times \mathbf{A} = 0$) the commutation relations of **K** reduce to those of **L**.

Consider a system for which the magnetic flux Φ is constant in time, and therefore does not produce at the particle a local electric field by means of Faraday's law of induction. Thus, with α independent of time, the Schrödinger equation

$$\varepsilon((-i/\hbar)\,\partial/\partial\varphi - \alpha)^2\Psi = i\hbar\,\partial\Psi/\partial t \tag{36}$$

readily admits to *two* stationary-state solutions with the following wave functions, energy, and angular-momentum eigenvalues.

(A) Solution I

$$\Psi_\mu^{\mathrm{I}}(\varphi, t) = \frac{1}{\sqrt{2\pi}}\exp[i(\mu + \alpha)\varphi]\exp(-iE_\mu^0 t/\hbar) \qquad (\mu = 0, \pm 1, \pm 2, \dots). \tag{37a}$$

Energy: $\qquad\qquad\qquad\qquad E_\mu^0 = \mu^2\varepsilon.$ $\qquad\qquad\qquad\qquad$ (37b)

Canonical angular momentum: $\quad l_\mu = (\mu + \alpha)\hbar.$ $\qquad\qquad\qquad$ (37c)

Kinetic angular momentum: $\qquad k_\mu = \mu\hbar.$ $\qquad\qquad\qquad\qquad$ (37d)

(B) Solution II

$$\Psi_\mu^{II}(\varphi, t) = \frac{1}{\sqrt{2\pi}} \exp(i\mu\varphi) \exp(-iE_\mu t/\hbar) \qquad (\mu = 0, \pm 1, \pm 2, \dots). \quad (38a)$$

Energy:
$$E_\mu = (\mu - \alpha)^2 \varepsilon. \tag{38b}$$

Canonical angular momentum: $\quad l_\mu = \mu\hbar.$ \hfill (38c)

Kinetic angular momentum: $\quad k_\mu = (\mu - \alpha)\hbar.$ \hfill (38d)

A cursory examination of the solutions shows that if the magnetic flux were quantized in units of the fluxon, then α would be integer-valued and the two solutions entirely equivalent; i.e., to each state of Solution I labeled by μ would correspond a state of Solution II with identical eigenvalues although different labels. However, the flux through a current ring (like the two-dimensional rotator) is not ordinarily quantized in which case α can take on a continuum of values, and the two sets of solutions are not equivalent. Solution I has the energy spectrum of the field-free planar rotator, an integer kinetic angular-momentum spectrum, and a flux-dependent canonical angular-momentum spectrum. For nonintegral values of α, the wave function is multiple-valued. Solution II, by contrast, is characterized by an integer-valued canonical angular-momentum spectrum, a flux-dependent kinetic angular-momentum spectrum, and a single-valued wave function. The property of single-valuedness was invoked earlier (1d) to obtain the wave functions and energy spectrum of the field-free rotator; without single-valuedness, the eigenvalue spectrum of the angular momentum (canonical and kinetic are here equivalent) would not be integer-valued. Does this mean one should discard Solution I as unphysical? On the other hand, Solution I (37a) has the form of (1.6a) which gives rise to the AB effect. What is one to make of this?

The physical content of these two sets of solutions may be understood by examining them as special cases of the general solution to the rotator in the presence of a time-dependent flux. This solution is given by

$$\psi(t) = \exp\left(-(i/\hbar)\int^t H(t')\,dt'\right)\psi(0) = \exp\left(-i\varepsilon\int^t [L_z - \alpha(t')]^2\,dt'\right)\psi(0),$$
$$\tag{39a}$$

where the first equality is the general solution to the time-dependent Schrödinger equation whenever the Hamiltonian at two different times commutes, i.e., $[H(t), H(t')] = 0$. If, at $t = 0$, the potential-free rotator is in an eigenstate of angular momentum with quantum number μ, then, by relation (1.6a), the state of the rotator in the presence of a static vector potential $\mathbf{A}_0$ with sole tangential component $A_{0\varphi}(r) = \Phi(0)/2\pi r$ is represented by the wave function

$$\psi_u(0) = \frac{1}{\sqrt{2\pi}} \exp(i(\mu + \alpha_0)\varphi), \tag{39b}$$

where evaluation of the phase integral in (1.6a) has led to

$$(e/\hbar c) \int^{\mathbf{x}} \mathbf{A}_0 \cdot \mathbf{ds} = (e/\hbar c) \int^{\varphi} (\Phi(0)/2\pi r) r \, d\varphi' = (\Phi(0)/\Phi_0)\varphi = \alpha_0 \varphi. \tag{39c}$$

Substitution of relation (39b) into (39a) then yields the solution

$$\psi_u(\varphi, t) = \frac{1}{\sqrt{2\pi}} \exp(i(\mu + \alpha_0)\varphi) \exp\left(-i\varepsilon \int^t [\mu + \alpha_0 - \alpha(t')]^2 \, dt'\right). \tag{39d}$$

Consider two cases for the time dependence of $\alpha(t)$.

In the first case, the flux $\Phi(0)$ is null for $t < 0$ and is subsequently brought to the constant value Φ at $t = 0_+$. Then $\alpha_0 = 0$ for $t \le 0$, and $\alpha(t) = \alpha = \text{constant}$ for $t > 0$. By Faraday's law of induction the initiation of the magnetic flux generates an electric field, whose tangential (and only) component is given by

$$E_\varphi(t) = -\frac{1}{2\pi rc} \partial\Phi/\partial t, \tag{40a}$$

which exerts a torque on the rotating particle

$$\tau_z = erE_\varphi = -(e/2\pi) \partial\Phi/\partial t \tag{40b}$$

along the axis of rotation. The torque changes the initial angular momentum $L_z = \mu\hbar$ by the amount

$$\Delta L_z = \int_0^\infty \tau_z \, dt = -e\Phi/2\pi c = -\alpha\hbar, \tag{40c}$$

thereby giving the state a final mechanical angular momentum

$$K_z = L_z + \Delta L_z = (\mu - \alpha)\hbar. \tag{40d}$$

In a similar way the induced electric field also does work on the particle at a rate

$$dW/dt = \omega_{rot}\tau_z = (L_z/mr^2)\tau_z = -2\varepsilon(\mu - \alpha) \, \partial\alpha/\partial t, \tag{41a}$$

leading to a change in energy by the amount

$$\Delta E = \int (dW/dt) \, dt = -2\varepsilon\alpha(\mu - \tfrac{1}{2}\alpha). \tag{41b}$$

The final energy is then

$$E_\mu = E_\mu^0 + \Delta E = (\mu - \alpha)^2\varepsilon. \tag{41c}$$

The properties of Solution II can therefore be thought to derive from the local interaction of the particle with an induced electric field over the period of initiation of the magnetic flux (which, in fact, can be effected at an entirely arbitrary rate).

In the second case we allow the flux through the circular space of the rotator to remain at its initial value throughout the existence of the rotator system. Then $\alpha_0 = \alpha(t) = \alpha = $ constant, and it is seen that the resulting wave function is that of Solution I. As there is no induced electric field to exert torques and do work on the charged particle, one would expect—as indeed is the case—that there is no effect of the flux on the characteristic energy and mechanical angular momentum of the system. Nevertheless, there is a purely quantum mechanical influence, a bound-state AB effect, reflected in the flux-dependent shift of the canonical angular-momentum eigenvalue spectrum.

It is worth noting the general significance of past history on the present state of a quantum system as illustrated by the example of the two-dimensional rotator in a vector potential field. Although both solutions represent a rotator threaded by a "constant" magnetic flux—constant, that is, from the perspective of one examining the system long after its formation—the manner of formation has played a seminal role. The properties of the orbiting particle, which at some point in its past experienced the initiation of the magnetic flux, are different from an apparently identical particle which began its orbital motion about an already existing tube of flux. There is nothing in this perspective that violates quantum mechanical principles although the present situation is not frequently encountered. After all, is not a 2S state the same for all hydrogen atoms whether it was the end result of an excitation from the 1S ground state or a decay transition from a 3P excited state? In most cases the answer is undoubtedly "yes," but there are circumstances, such as just illustrated, where the mode of origination of a quantum system leaves a distinct legacy.

What about the admissibility of multiple-valued wave functions? This has been a long-debated and somewhat thorny issue. Some have argued that single-valuedness is a necessary criterion for solutions to be physically meaningful [21, 24]; others have taken the position that multiple-valued wave functions should not be rejected *a priori* [25, 26]. Whereas the imposition of single-valuedness as a criterion for an electromagnetic or gravitational field to be physically meaningful is equivalent to requiring that the classical forces acting on a particle be uniquely specified, the same criterion applied to a quantum field, such as the quantum mechanical wave function, is not so transparent and readily motivated; it is a *bilinear* product of wave functions (like the probability density), and not the wave function itself, that is directly related to experimental observables.

In a simply connected space, such as that of the planar rotator in the absence of magnetic flux, specification of single-valuedness ensures against spurious solutions introduced by the arbitrary choice of a particularly convenient coordinate system with the polar axis through the loop. An alternative choice of polar axis outside the loop leads to single-valued solutions only. The mathematical equivalence of the two descriptions reflects the topological circumstance that there is no unambiguous meaning to the "inside" of a closed loop in a simply connected space. Topologically, any such loop can be deformed continuously to a point. In the case of a nonsimply connected space, however, such as that of the planar rotator threaded by a tube of magnetic flux, there is a significant topological distinction between loops that thread once, twice, thrice, etc., about a hole in the space. Two loops that wind about the excluded region of space a different number of times cannot, because of the hole, be continuously deformed into one another; mathematically, they fall into different so-called homotopy classes.

The argument for rejection of multiple-valued wave functions has also been made on the grounds that they fail to regenerate the expected single-valued solutions upon adiabatic extinction of the magnetic flux. Such an argument would not be relevant, however, to a composite system for which the particle experiences an invariable flux beyond the control of the experimenter.

Finally, it has often been asserted that the two-dimensional system of a particle orbiting an infinitely long tube of flux is in reality just an idealization of a three-dimensional system with high, but finite, potential barrier and finite return flux. Then, there would again no longer be a distinction between paths that wind about a designated origin and paths that do not (for all such paths could be continuously deformed into one another), and wave functions should once more be single-valued. This point of view entirely sidesteps the original problem by converting it to a three-dimensional one. The cogency of the premise will not be discussed here further except to say that the two-dimensional system as introduced is a conceptually valid one, and diverse theoretical studies have shown that the physics of two dimensions can differ qualitatively from the physics of three.

2.5. Fermions, Bosons, and Things In-Between

If real systems analogous to the planar rotator threaded by a temporally invariant magnetic flux should exist in nature, these systems would manifest curious physical properties indeed. As discussed earlier, the wave function of a particle rotated, let us say by an angle θ, undergoes a unitary transformation generated by the canonical angular-momentum operator

$$\psi(\varphi + \theta) = \exp(-iL_z\theta/\hbar)\psi(\varphi). \tag{42a}$$

Thus, an eigenfunction of L_z with the flux-dependent eigenvalue spectrum (37c) subjected to an integral number of complete rotations—equivalent to the identity operation in a simply connected space—takes the form

$$\psi(\varphi + 2\pi n) = \exp(-i2\pi n\Phi/\Phi_0)\psi(\varphi). \tag{42b}$$

In the absence of magnetic flux, or more generally for a magnetic flux quantized in integral units of the fluxon, the initial and rotated wave functions are identical, and the system behaves like a scalar under rotation. For Φ/Φ_0 equal to a half-integral number of fluxons, however, the phase factor equals -1, and the wave function rotates like a spinor. Spinors ordinarily provide a mathematical description of fermions, particles (like the electron or neutron) with half odd integer values of intrinsic angular momentum or spin. In the present circumstances, the actual spin of the system is irrelevant, and therefore even a spinless charged boson could, according to (42b), exhibit the rotation properties of a fermion [7, 8]. Of course, the system must not be thought of as the particle alone, but rather as the composite bound system of a particle and magnetic flux tube. If the magnetic flux is arbitrary, then the phase factor in (42b) is neither $+1$ nor -1, but is in general a complex number.

The connection between the angular momentum (and therefore the rotational properties) of individual particles and the quantum statistical behavior of many such identical particles has never been particularly transparent as quantum analyses go, and, in fact, actually lies outside quantum mechanics proper, having its origin in arcane propositions relating to the microscopic causality of relativistic quantum fields [27]. It is indeed amazing, as some authors have pointed out, that "such a deep connection, essential to the stability of matter in circumstances apparently very remote from the relativistic domain, does require these concepts. We do not know any alternative basis for it" [28]. The end result, however, can be simply stated and is widely known—namely, all matter should fall into one of two categories of particles: bosons and fermions. The wave function of a multiboson system is unaltered by the exchange of location of any two particles, one consequence of which is that there is no upper limit to the number of bosons that can occupy a given quantum state. In contrast, the wave function of a multifermion system undergoes a change of sign for each pair of particles exchanged, and not more than one fermion within the system can occupy a specified quantum state. From a group-theoretical perspective the wave function of n bosons or n fermions is, respectively, the fully symmetric or fully antisymmetric irreducible representation of the permutation group S_n, the other representations [29] of S_n apparently not being used by Nature in this regard.

However, theoretical examination of composites like the planar rotator bound to a tube of magnetic flux indicates that the exchange of two identical composites multiplies the wave function of the total system by a complex phase factor like that in (42b). Since the phase in this factor can have any value, the composite of particle and flux has been termed an "anyon" [9]. The study of anyons raises the fascinating question of whether Fermi–Dirac and Bose–Einstein statistics are actually special cases of a more general quantum statistics and, indeed, a general theory of the quantum statistics of two-dimensional systems has been developed [30].

Do anyons—or anything closely resembling them in the three-dimensional world—actually exist? The answer is quite possibly an affirmative one, although definitive proof has yet to be provided. Nevertheless, their existence has been postulated in theories of diverse phenomena like the fractional quantum Hall effect (where the effective charge carriers appear to come in fractions of a single electron charge) and high-temperature super-conductivity [31].

Two objections are perhaps likely to arise at this point concerning the direct observability, even in principle, of the flux-dependent phase factor. First, how would one be able to know whether the bound particle has rotated around the flux tube or not? The canonical angular momentum and the angular coordinate φ (or, more precisely, harmonic functons of φ [32]) are effectively conjugate variables which, by the uncertainty principle cannot be known simultaneously with perfect accuracy. Since the states of the planar rotator are characterized by sharp (albeit flux-dependent) values of angular momentum, the wave function is delocalized over the entire available space—like a standing wave, rather than like a particle. Second, the phase factor is a global factor independent of the state of the system, and therefore vanishes when one calculates expectation values or transition probabilities. How can it then enter the mathematical expression for some observable quantity?

Although it is impossible to ascertain the angle of rotation of a particle in one of the planar rotator eigenstates, one can always conceive of a particle in a state described by a linear superposition of such eigenstates spanning a range of quantum numbers μ. The greater the uncertainty in angular momentum, the more localized will be the wave packet representing the state of the particle. Moreover, since each superposed eigenstate contributes the same flux-dependent phase factor under rotation, the rotationally trans-formed wave packet will have exactly the same form as expression (42b). With regard to the second question, it should be noted that under the conditions of a quantum interference experiment in which rotational path-ways about two magnetic flux tubes are available to the particle, the phase in (42b) would cease to be global, the experimental outcome then being sensitive to the *relative* phase of the two probability amplitudes.

To test the predicted influence of the isolated magnetic flux, it is not necessary to limit one's considerations to a strictly periodic system like the two-dimensional rotator. Consider instead a configuration such as that of Figure 2.5 in which a collimated beam of charged particles (with or without spin) is split into two coherent beams and made to circulate in orbits of equal radius and opposite sense about similar solenoids generating fluxes Φ_1, Φ_2, respectively [7]. As before, use of the classical terminology of "splitting a beam," should not disguise the intrinsically quantum nature of

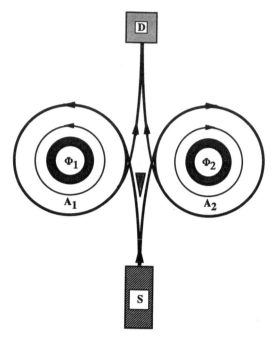

Figure 2.5. Split-beam electron interference experiment manifesting the influence of winding number in the AB effect. Depending on the value of the magnetic flux, a spinless charged particle can behave under rotation like a fermion. (Adapted from Silverman [7].)

the experiment; what is intended is that one particle at a time may propagate through the apparatus—just as in the interference experiments with an electron-microscope beam. It is therefore the two components of a *single-particle* wave packet that propagate around the two solenoids. If ever probed, one would find, of course, that there is but one particle and that it had taken one circular path or the other, but not both. To make particles circulate in the prescribed manner one could in principle employ uniform, time-independent background magnetic fields of equal strength and opposite orientation; these background fields would then contribute no net relative phase to the AB phase shift.

Upon recombination of the coherently split beam after n revolutions about one or the other solenoid, the wave function (for parallel tubes of flux) takes the form

$$\psi(\varphi + 2\pi n) = \frac{1}{\sqrt{2}} [\exp(-i2\pi n\Phi_1/\Phi_0)\psi_1(\varphi) + \exp(+i2\pi n\Phi_2/\Phi_0)\psi_2(\varphi)].$$

$$(43a)$$

For equal amplitudes ψ_1 and ψ_2, the state (43a) leads to a forward beam intensity

$$I(2\pi n) \propto |\psi(\varphi + 2\pi n)|^2 \propto \cos^2[2\pi n(\Phi_1 + \Phi_2)/\Phi_0]I_0, \qquad (43b)$$

where I_0 is the incident beam intensity. Thus, the predicted phase factor can be made to have experimental consequences in a split-beam quantum interference configuration. Although there is no direct contact between the particles of the beam and the magnetic fields within the two solenoids, the forward beam intensity reveals the number of times a particle has circulated around one of the solenoids (although not, of course, which one).

The quantum number n in (42b) and ensuing relations is related to the topological concept of winding number which plays a fundamental role in the study of the connectivity of spaces. Paths through a space can be defined in such a way that they form a group known as the homotopy group [33]. One might think that there would ordinarily be an infinite number of different paths, but from the perspective of group theory this is not necessarily so. There are as many distinct homotopic classes of paths—paths that cannot be continuously deformed into one another (i.e., without cutting and pasting)—as there are distinct group operations. In a three-dimensional space without holes—representable by the unit sphere—there are only two classes of paths. All paths that begin at the origin, and cut the surface of the sphere an even number of times, can be deformed to a single point at the origin; conversely, all paths that cut the surface an odd number of times can be deformed into a simple closed path between the origin and a single point on the surface. Thus, there are only two homotopy classes for a three-dimensional sphere: even and odd. Surprisingly, the smaller two-dimensional space of an annulus is homotopically the richer. Paths that have the same endpoints and the same winding number can be deformed into one another and therefore belong to the same class; they are homotopically equivalent. The homotopy group of an annulus contains an infinite number of classes.

In the two-slit AB experiment of Chapter 1, only the classical paths available to the electron—those that pass above and below the solenoid—were included in the superposition of probability amplitudes (see (10a)). Is it conceivable, however, that every once in a while an electron in the beam makes one or more full loops about the solenoid before propagating to the detector? The existence of such nonclassical paths of higher winding number can be justified by theory [34], but the amplitudes for these processes must be very weak, for no evidence of their contribution has yet been discerned in the resulting interference patterns.

2.6. Quantum Interference in a Metal Ring

Although electron beam experiments in vacuum have so far exhibited no distinguishable quantum effects attributable to particle winding, such effects *have* been seen in a two-dimensional system that at first thought would

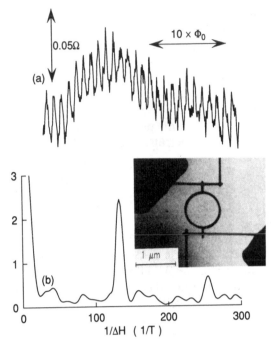

Figure 2.6. Electron interference in a mesoscopic normal metal ring shown in the insert. The resistance as a function of magnetic field (Curve a) displays oscillations with period Φ_0 attributable to the **AB** effect for electron trajectories between the entrance and exit terminals. The Fourier transform (Curve b) also reveals an oscillation period of $\Phi_0/2$ for cyclic electron trajectories beginning and ending at the point of entry. (Courtesy of S. Webb, University of North Carolina.)

appear most unlikely to display any quantum interference at all. In recent years experimental investigations of very small, normal (i.e., not super-conducting) metal rings have revealed surprising quantum behavior in total contrast to that anticipated from the classical theory of metals (or even from a number of incautious applications of quantum theory) [35].

Figure 2.6 shows a representative sample, a gold ring of approximately 1 μm in diameter with wire thickness of a few hundredths of a micron. The ring, which contains about 10^8 atoms, is said to be of mesoscopic size, i.e., a scale between atomic and macroscopic dimensions. (Cells of the human body are roughly 5–20 μm in size.) The tiny ring is part of a circuit; electrons are introduced at one terminal, flow through the ring, and exit at the diametrically opposite terminal.

Except for topology, there is seemingly little in common between a metal ring with its lattice of some millions of positive ions and ambient sea of bound and conduction electrons, and the single electron circulating in a space devoid of other matter. Indeed, the latter system does not have a resistance. By contrast, a conduction electron in a normal disordered metal ring does not propagate freely, but diffuses through the metal in the manner of a random walk, undergoing collisions with impurity atoms and lattice defects. The mean free path of such an electron, typically on the order of 100 Å (10^{-2} μm), is about one hundred times smaller than the sample length. As a consequence of the frequent collisions, it is intuitively reasonable to suppose that the phase of the wave function of the electron at the exit terminal should be essentially unrelated to the phase at the entrance. In other words, there should be no electron coherence across the ring, so that the probability amplitudes for passage between entrance and exit by clockwise or counterclockwise paths about the hole could not interfere. Thus, one would not expect to observe the AB effect in a mesoscopic metal ring subjected to a magnetic field through the central hole.

Contrary to such expectations, the metal ring *does* exhibit the AB effect, as revealed by oscillations (with period $\Phi_0 = hc/e$) in the resistance of the ring as a function of the magnetic field, as also shown in Figure 2.6. Strictly speaking, the conditions for the AB effect do not hold in these experiments, for the magnetic field penetrates the entire ring and not solely the central hole. The unbound electrons are therefore subject to the Lorentz force, but this is *not* the source of the oscillations (although field penetration of the ring has important consequences as will be discussed shortly). The main point at present which requires explanation is the surprising survival of electron coherence. How can a particle that has undergone hundreds or thousands of scatterings exhibit self-interference?

The solution to the enigma is that *elastic* collisions, which predominate in a reasonably pure metal at low temperature (below 1 K), do not cause loss of phase memory; the electron coherence is destroyed only by inelastic collisions. Consequently, the coherence length of the electron is not to be identified with the total mean free path, but rather only with the mean distance between inelastic scatterings or other phase-randomizing events, which can attain values of a few microns—i.e., a size larger than the ring. Until the recent experiments on mesoscopic rings, prevailing opinion held that *all* scattering would destroy the electron coherence. This belief is now known to be false.

The idea of coherence—which ordinarily refers to an ensemble of similarly prepared systems—must be interpreted a little differently in the present case, since the phase of the exiting electron wave function is a function of the random path through the metal. One, therefore, cannot prepare the quantum states of conduction electrons in the ring in such a way as to control the phase of exiting electrons. Nevertheless, as elastic collisions are reversible, the phase is well determined for any given path, and therefore all electrons that follow the identical path will exhibit self-interference with the same relative phase. The conductance of a mesoscopic metal ring can vary markedly from one sample to another, the average over many such rings yielding the bulk value deducible theoretically by a quantum mechanical ensemble average. An ensemble average, however, does not in general apply to an individual ring.

As a consequence of mesoscopic coherence, there is a nonvanishing interference between probability amplitudes of a carrier that has diffused from the entrance to the exit terminal by passing (once) to one side or the other of the hole. In addition to a random geometrical phase shift acquired by propagation and scattering (and therefore sensitive to the distribution of impurities in the metal), the interference term acquires, as in the AB effect with free particles, a supplementary magnetic phase shift $(2\pi\Phi/\Phi_0)$ determined by the amount of magnetic flux enclosed by the path. However, because the magnetic field in the mesoscopic ring experiments was not confined to the hole, but uniformly permeated the entire annulus, different pairs of paths can enclose different values of flux. Thus, the AB phase shift—like the geometrical phase shift—is also path-dependent. Nevertheless, the fluctuation in enclosed flux is smaller than the flux through the central hole by approximately the ratio of the area of the annulus to the area of the hole, which, by experimental design, is a very small number. Because of the path sensitivity of both geometric and magnetic phase shifts, the variation of resistance with magnetic field displays an aperiodic random background upon which is superposed—as the signature of the AB effect—an oscillatory fine structure with period Φ_0.

The periodicity of the oscillations can be accurately established by electronically determining the Fourier transform of the magneto-resistance data. With judicious choice of ring geometry, however, the Fourier transform shows an additional peak at a magnetic field corresponding to an oscillation period $\Phi_0/2$ as shown again in Figure 2.6. It is perhaps no surprise, in view of the preceding discussion on winding, that an oscillation with phase angle $\Phi/(\Phi_0/2) = 2\Phi/\Phi_0$ is attributable to interference involving two paths—one clockwise, the other counterclockwise—making one complete revolution about the ring. (Since electrons following a 360° path return to their point of injection and do not leave the ring, they therefore contribute to a diminution in conductance or increase in resistance.) The two counter-revolving paths need not be identical, in which case they enclose different values of flux, and the contribution of all such pairs of paths within the ring is again a statistical one. Theoretically, the total electrical resistance of the ring as a function of magnetic field B can be expressed in the form

$$R(B) = R_b + \sum_{n=1} R_n \cos[\alpha_n + 2\pi(n\Phi/\Phi_0)], \tag{44}$$

where R_b is the aperiodic background comprising the classical resistance (including a magneto-resistance term proportional to B^2) and the zero-frequency part of the AB effect, and R_n and α_n (for integer $n = 1, 2, \ldots$) are random functions of B approximately inversely proportional to the winding number n and varying over a domain of B proportional to the area of the annulus.

There is one circumstance, however, where the AB effect in a mesoscopic disordered metal is *not* a statistical one—i.e., is totally insensitive to the distribution of impurities along the path. This is the case where the magnetic field is confined to the central hole, and contributing paths through the metal are precisely time-reversed images of one another. In other words, if one path through the metal involves electron scattering at impurity sites $1, 2, 3, \ldots, N$, then the other path involves the same scattering events in reverse order, $N, \ldots, 3, 2, 1$. Such pairs of time-reversed trajectories can *only* occur for complete revolutions about the ring—and therefore give rise to magneto-resistance oscillations with maximum period $\Phi_0/2$.

Ironically, it was just this condition that prevailed in the first observation of the AB effect in a normal metal structure—with the configuration of a cylinder, not a two-dimensional ring [36]. The sample was a quartz fiber about 1 cm in length and 1 μm in diameter upon which a metal film had been deposited. In stark contrast to the features of the AB effect in a ring described above, the variation of sample resistance with magnetic field yielded a smoothly oscillatory function with period $\Phi_0/2$ only. There was no contribution at the fundamental period Φ_0. Here is an example in which the AB effect derives exclusively from the occurrence of winding.

One might have expected that mesoscopic cylinders and rings—which, after all, are topologically equivalent—should exhibit similar magneto-resistance behavior. Why, then, were no AB oscillations observed at the fundamental period Φ_0 in the cylinder? A cylinder may be thought of as a parallel stack of mesoscopic rings. If the length of the stack is much greater than the coherence length of a ring, the AB oscillations with period Φ_0 from different rings (i.e., from segments of the cylinder separated by more than the coherence length) are uncorrelated and average out, leaving only a contribution from the special subset of time-reversed paths. In essence, the three-dimensional cylindrical geometry effects an ensemble average of mesoscopic rings [37].

Nevertheless, whether in a single ring or cylindrical stack, the contribution of time-reversed paths to the AB effect is rapidly extinguished when magnetic flux penetrates the metal to an extent on the order of Φ_0. Different pairs of paths then enclose different values of flux, and the resulting oscillations are no longer in phase.

References

[1] The diagrammatic representation of a "conjugated" molecule like benzene,

C_6H_6, shows alternating single and double bonds, but, as a result of the pi electron

delocalization, any pair of adjacent carbon atoms in ·the ring is indistinguishable from any other pair.

[2] K.B. Wiberg, *Physical Organic Chemistry* (Wiley, New York, 1964), p. 9.

[3] M.P. Silverman, Broken Symmetry of the Charged Planar Rotator in Electric and Magnetic Fields, *Amer. J. Phys.*, **49**, 871 (1981).

[4] M.P. Silverman, Angular Momentum and Rotational Properties of a Charged Particle Orbiting a Magnetic Flux Tube., *Fundamental Questions in Quantum Mechanics*, Edited by L. Roth and A. Inomata (Gordon & Breach, New York, 1986), pp. 177–190.

[5] D.S. Carlstone, Factorization Types and SU(1, 1) *Amer. J. Phys.*, **40**, 1459–1468 (1972).

[6] S. Flügge, *Practical Quantum Mechanics I* (Springer-Verlag, New York, 1971), pp. 110–112.

[7] M.P. Silverman, Experimental Consequences of Proposed Angular Momentum Spectra for a Charged Spinless Particle in the Presence of Long-Range Magnetic Flux, *Phys. Rev. Lett.*, **51**, 1927 (1983).

[8] M.P. Silverman, Quantum Interference Test of the Fermionic Rotation Properties of a Charged Boson–Magnetic-Flux-Tube Composite, *Phys. Rev. D*, **29**, 2404 (1984).

[9] F. Wilczek, Magnetic Flux, Angular Momentum, and Statistics, *Phys. Rev. Lett.*, **48**, 1144 (1982); Quantum Mechanics of Fractional-Spin Particles, *Phys. Rev. Lett.*, **49**, 957 (1982).

[10] R. Jackiw and A.N. Redlich, Two-Dimensional Angular Momentum in the Presence of Long-Range Magnetic Flux, *Phys. Rev. Lett.*, **50**, 555 (1983).

[11] M.P. Silverman, Exact Spectrum of the Two-Dimensional Rigid Rotator in External Fields, I. Stark Effect, *Phys. Rev. A*, **24**, 339 (1981).

[12] See, for example, L.I. Schiff, *Quantum Mechanics*, 3rd edn. (McGraw-Hill, New York, 1968), Chapter 8, pp. 244–255, 263–268.

[13] N.W. McLachlan, *Theory and Application of Mathieu Functions* (Dover, New York, 1964), p. 10.

[14] L. Brillouin, *Wave Propagation in Periodic Structures* (Dover, New York, 1953), Chapter 8.

[15] E.T. Whittaker and G.N. Watson, *A Course of Modern Analysis* (Cambridge University Press, London, 1969), p. 417.

[16] E.P. Wigner, *Group Theory* (Academic Press, New York, 1959), p. 144.

[17] See, for example, I.V. Schensted, *A Course on the Application of Group Theory to Quantum Mechanics* (Neo Press, Maine, 1976), p. 119.

[18] A. Messiah, *Quantum Mechanics II* (Wiley, New York, 1962), pp. 696 and 709.

[19] H.V. McIntosh, On Accidental Degeneracy in Classical and Quantum Mechanics, *Amer. J. Phys.* **27**, 620–625 (1959).

[20] C. Cohen-Tannoudji, B. Diu, and F. Laloë, *Quantum Mechanics*, Vol. I (Wiley, New York, 1977), p. 136.

[21] D. Bohm and B.J. Hiley, On the Aharonov–Bohm Effect, *Nuovo Cimento*, **52A**, 295–308 (1979).

[22] K-H. Yang, Gauge Transformations and Quantum Mechanics I. Gauge Invariant Interpretation of Quantum Mechanics, *Ann. Phys. (N.Y.)*, **101**, 62 (1976).

[23] M.P. Silverman, Rotation of a Spinless Particle in the Presence of an Electromagnetic Potential, *Lett. Nuovo Cimento*, **41**, 509–512 (1984).

[24] E. Merzbacher, Single Valuedness of Wave Functions, *Amer. J. Phys.*, **30**, 237–247 (1962).

[25] J.M. Blatt and V.F. Weisskopf, *Theoretical Nuclear Physics* (Wiley, New York, 1952), pp. 783, 787.

[26] M.P. Silverman, On the Use of Multiple-Valued Wave Functions in the Analysis of the Aharonov–Bohm Effect, *Lett. Nuovo Cimento*, **42**, 376–378 (1985).

[27] J.D. Bjorken and S.D. Drell, *Relativistic Quantum Fields* (McGraw-Hill, New York, 1965), pp. 170–172; see, too, the following reference.

[28] C. Itzykson and J.-B. Zuber, *Quantum Field Theory* (McGraw-Hill, New York, 1980), pp. 149–151.

[29] The total number of classes $r(n)$ of the permutation group S_n is equal to the number of ways to partition n into a set of positive integers that sum to n. This is given by the coefficient of x^n in the formal power series expansion of the Euler generating function. See J.S. Lomont, *Applications of Finite Groups* (Academic Press, New York, 1959), p. 259.

[30] Y.-S. Wu, General Theory for Quantum Statistics in Two Dimensions, *Phys. Rev. Lett.*, **52**, 2103 (1984).

[31] B. Halperin, J. March-Russell, and F. Wilczek, Consequences of Time-Reversal Symmetry Violation in Models of High-T_C Superconductors, *Phys. Rev. B*, **40**, 8726 (1989).

[32] P. Carruthers and M.M. Nieto, Phase and Angle Variables in Quantum Mechanics, *Rev. Mod. Phys.*, **40**, 411 (1968).

[33] R. Gilmore, *Lie Groups, Lie Algebras, and Some of Their Applications* (Wiley–Interscience, New York, 1974), pp. 129–134.

[34] C. Bernido and A. Inomata, Topological Shifts in the Aharonov–Bohm Effect, *Phys. Lett.*, **77A**, 394 (1980).

[35] See, for example, (a) S. Washburn and R.A. Webb, Aharonov–Bohm Effect in Normal Metal Quantum Coherence and Transport, *Adv. in Phys.*, **35**, 375–422 (1986); (b) R.A. Webb and S. Washburn, Quantum Interference Fluctuations in Disordered Metals, *Physics Today*, **41**, 46–53 (Dec. 1988); (c) S. Washburn, Conductance Fluctuations in Loops of Gold, *Amer. J. Phys.* **57**, 1069–1078 (1989).

[36] D. Yu. Sharvin and Yu. V. Sharvin, Magnetic Flux Quantization in a Cylindrical Film of a Normal Metal, *JETP Lett.* **34**, 272–275 (1981).

[37] A.D. Stone and Y. Imry, Periodicity of the Aharonov–Bohm Effect in Normal-Metal Rings, *Phys. Rev. Lett.*, **56**, 189 (1986).

CHAPTER 3

Interferometry of Correlated Particles

And the only tune I hear
Is the sound of the wind
As it blows through the town
Weave and spin; weave and spin.

Aragon Mill
Si Kahn

3.1. Ghostly Correlations: Wave and Spin

In a 1935 paper [1] that has since become a classic in the literature regarding conceptual implications of quantum mechanics, Einstein and his colleagues Boris Podolsky and Nathan Rosen (to be designated EPR) raised in one of its starkest forms the issue of nonlocality—that is, the occurrence of interactions instantaneously at a distance in violation of physicists' intuitive sense of cause and effect as embodied in the principles of special relativity. Actually, Einstein's principal focus of concern was the completeness of quantum mechanics as a self-consistent theory of individual particles (as opposed to a purely statistical theory of ensembles of particles), but the *Gedankenexperiment* proposed by EPR illustrated what many physicists throughout the ensuing years have considered to be one of the strangest features of quantum mechanics.

In its barest essentials, the EPR experiment concerns the correlation of coordinates and momenta of two particles that have interacted at some time in the past and then separated to such an extent that for all practical purposes (from the perspective of classical physics) they would appear to be distinctly independent systems at the time a measurement was to be performed on them. Although it is not possible to measure both the position and momentum of each particle simultaneously—to do so would violate the uncertainty principle—one could in principle, according to EPR, measure one of these variables for one particle of the pair and, *without in any way disturbing the state of the second particle*, deduce the corresponding variable

59

with 100% certainty. However, since the choice of whether to measure coordinate or momentum is a decision to be made by the experimenter—and this decision can even be made *after* the two particles have separated—the remote unprobed particle cannot "know" *which* measurement was affected on the examined particle. A different experimental configuration is required to measure position than to measure momentum, and, depending upon which configuration is employed, the unprobed particle would be expected to manifest sharp values of either one or the other—but not both—of two canonically conjugate variables. Can the fact that a measurement was made on one particle be transmitted instantaneously to the other? If not, then do the particles have well-defined positions and momenta even though both properties are not simultaneously measurable? This enigma of completeness versus locality constitutes in part what has been termed the EPR paradox.

The quantitative features of the one-dimensional model system analyzed by EPR can be briefly summarized in the following way (as first shown by Niels Bohr). From the momenta (p_1, p_2) and coordinates (q_1, q_2) of the two separated particles, which satisfy the usual quantum commutation conditions (with $i, j = 1, 2$)

$$[q_i, q_j] = 0, \qquad [p_i, p_j] = 0, \tag{1a}$$

$$[q_i, p_j] = i\hbar\delta_{ij}, \tag{1b}$$

one can define new pairs of conjugate variables (Q_1, P_1) and (Q_2, P_2) by means of a rotational transformation with parameter θ

$$\begin{pmatrix} Q_1 \\ Q_2 \end{pmatrix} = \begin{pmatrix} \cos\theta & \sin\theta \\ -\sin\theta & \cos\theta \end{pmatrix} \begin{pmatrix} q_1 \\ q_2 \end{pmatrix}, \tag{1c}$$

$$\begin{pmatrix} P_1 \\ P_2 \end{pmatrix} = \begin{pmatrix} \cos\theta & \sin\theta \\ -\sin\theta & \cos\theta \end{pmatrix} \begin{pmatrix} p_1 \\ p_2 \end{pmatrix}, \tag{1d}$$

The transformed variables satisfy commutation rules identical in form to (1a, b). Although it is not possible to assign definite numerical values to both Q_1 and P_1 (since they do not commute), one could, however, prepare the two-particle system in a state such that Q_1 and P_2 (which *do* commute) have known, sharp values. Then, since

$$Q_1 = q_1 \cos\theta + q_2 \sin\theta; \qquad P_2 = -p_1 \sin\theta + p_2 \cos\theta; \tag{1e}$$

a measurement of either q_1 or p_1 will allow one to predict, respectively, the corresponding quantity q_2 or p_2.

Although recognizably arbitrary, EPR adapted as a reasonable definition of reality the criterion that: "If, without in any way disturbing a system, we can predict with certainty (i.e., with probability equal to unity) the value of a physical quantity, then there exists an element of physical reality corresponding to this physical quantity." By this criterion, then, the coordinate and momentum of particle 2 must be real—since it can be predicted with certainty by measurements made on distant particle 1. Yet, according to quantum mechanics, a complete description of the system does not permit simultaneous knowledge of the coordinate and momentum of a particle.

Thus, quantum mechanics is incomplete, or the properties of a particle have no physical reality until measured, an implication anathema to any theory that purports to make objective sense of the world according to EPR.

Published replies to the EPR paper followed quickly from Bohr [2] and many others. In fact, as soon as the EPR paper was published, Einstein personally received a large number of letters from physicists pointing out to him where the argument failed. He found it considerably amusing that "while all the scientists were quite positive that the argument was wrong, they all gave different reasons for their belief!" [3]. In recent times entire conferences devoted to the conceptual foundations of quantum theory testify still to the undiminished fascination that many physicists and laymen alike have with the issues raised by EPR. Numerous experimental tests of the EPR correlations have since been performed, although not by measuring the position and momentum of massive particles, but rather the polarization of correlated photons. As these experiments have been frequently discussed elsewhere in the literature [4], I will not remark upon them further, but consider instead the novel features of proposed experiments with *charged* particles which interact in ways for which no optical analogues exist.

Before doing this, however, it is necessary to consider a second type of particle correlation that finds its origin, not in the "entanglement" (to use a term of Schrödinger) of wave functions resulting from special state preparation as in the EPR paradox, but rather as an indirect consequence of spin through the spin-statistics connection discussed earlier. In the mid-1950s two astronomers, R. Hanbury Brown and R.Q. Twiss (to be designated HBT) developed a new type of interferometer whose underlying explanation was eventually to have as profound an impact on quantum physics—in particular quantum optics—as did the EPR paradox and the AB effect. Known as an intensity interferometer, the apparatus operated by correlating (i.e., multiplying together and time-averaging) the output currents from two photodetectors illuminated by light from a thermal source such as a star (Figure 3.1). Since it is the light *intensity* to which each detector responds and to which the output current is proportional, the resulting oscillatory correlation as a function of detector separation was regarded by many as highly surprising. All physicists know that wave *amplitudes*, not intensities, interfere. HBT took pains to point out that the phenomenon did not actually involve the interference of light intensities and could, in fact, be happily understood by a radio engineer within the framework of the classical wave theory of radiation. Nevertheless, physicists probing the quantum implications of the HBT effect were even more surprised, if not altogether incredulous, to learn that photons, emitted apparently randomly from a thermal source, were correlated in their arrivals at the two detectors [5]. In the charming description of Hanbury Brown, later reminiscing about this period [6]:

Now to a surprising number of people, this idea seemed not only heretical but patently absurd and they told us so in person, by letter, in publications, and by actually doing experiments which claimed to show that we were wrong. At the most

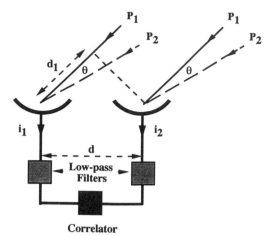

Figure 3.1. Hanbury Brown–Twiss (HBT) inensity interferometer. Wave fronts issuing from two points (P_1, P_2) of an extended source are received at two photodetectors whose output currents are first filtered to let pass low-frequency harmonics and then correlated. d_1 is the greater distance traveled by the wave front from P_1 to detector 1 than to detector 2. (The analogous interval d_2 for the wave front from P_2 is not shown.)

basic level they asked how, if photons are emitted at random in a thermal source, can they appear in pairs at two detectors. At a more sophisticated level the enraged physicist would brandish some sacred text, usually by Heitler, and point out that ... our analysis was invalidated by the uncertainty relation ....

Unfolding developments were to show, however, that HBT were not only *not* wrong, but their work—although originally conceived for the purpose of measuring stellar angular diameters—actually laid the experimental foundation for what has become modern quantum optics and provided new methods for fundamental tests of quantum mechanics.

A heuristic understanding of the intensity interferometer, schematically diagrammed in Figure 3.1, can be gained by examining the superposition at each detector of the classical electromagnetic waves emitted from two different locations in an extended optical source. Of the broad range of frequencies radiated by a thermal source, consider just two Fourier components of the same linear polarization, $E_1 \sin(\omega_1 t + \varphi_1)$ from point P_1 and $E_2 \sin(\omega_2 t + \varphi_2)$ from point P_2, that reach detector D_1. The phases φ_1 and φ_2 in the argument vary randomly from one emitted wave front to another. The instantaneous intensity at D_1, to which the output current i_1 is proportional (with detector proportionality constant K_1) then takes the form

$$i_1 = K_1[E_1 \sin(\omega_1 t + \varphi_1) + E_2 \sin(\omega_2 t + \varphi_2)]^2. \tag{2a}$$

A similar expression

$$i_2 = K_2[E_1 \sin(\omega_1(t + d_1/c) + \varphi_1) + E_2 \sin(\omega_2(t + d_2/c) + \varphi_2)]^2, \quad (2b)$$

gives the output photocurrent of detector D_2 where d_i ($i = 1, 2$) is the difference in optical path lengths between source point i and the two detectors. By computing the squares in (2a, b) and using trigonometric identities, one finds that the resulting photocurrents comprise Fourier components with frequencies ω_1, ω_2, $2\omega_1$, $2\omega_2$, $(\omega_1 + \omega_2)$, and $(\omega_1 - \omega_2)$. A low-frequency filter (e.g., with pass-band of approximately 1–100 MHz) that allows only the Fourier component at the difference frequency $(\omega_1 - \omega_2)$ to pass to the correlator gives rise to the following photocurrents from D_1 and D_2:

$$i_1 = K_1 E_1 E_2 \cos((\omega_1 - \omega_2)t + (\varphi_1 - \varphi_2)), \quad (2c)$$

$$i_2 = K_2 E_1 E_2 \cos((\omega_1 - \omega_2)t + (\varphi_1 - \varphi_2) + (\omega_1 d_1 - \omega_2 d_2)/c). \quad (2d)$$

Although the initial phases are random functions of time, the *difference* in these phases appears in the expressions for *both* photocurrents. These two photocurrents (and therefore the incident light intensities) are clearly correlated since at any instant they have the same frequency and differ in phase only by a constant term for a given interferometer configuration. Upon multiplying relations (2c, d) and time-averaging over one or more periods, one obtains the correlation function

$$\langle i_1(t) i_2(t) \rangle \approx K_1 K_2 E_1^2 E_2^2 \cos[(\omega/c)(d_1 - d_2)], \quad (2e)$$

where within the narrow pass-band it is adequate to set $\omega = \omega_1 \approx \omega_2$. A complete analysis of the intensity interferometer, which we do not need here, would require that (2e) be integrated over all pairs of points on the source and that all appropriate Fourier components be included. For our present purposes, however, the above expression is sufficient to show that the origin of the intensity "interference" can be readily understood on the basis of amplitude interference in classical wave theory.

It should be noted that technically HBT did not measure directly a correlation in the output photocurrents, but rather a correlation in the *fluctuations* of these currents. Representing each current as the sum of a stationary average term and a fluctuating term

$$i_k(t) = \bar{i}_k + \Delta i_k(t) \qquad (k = 1, 2), \quad (3a)$$

one can express the correlation in the fluctuations of the two beams by means of the so-called second-order correlation function (i.e., second order in the intensities or fourth order in the amplitudes)

$$g_{12}^{(2)} = \frac{\langle i_1(t_1) i_2(t_2) \rangle}{\bar{i}_1 \bar{i}_2} = 1 + \frac{\langle \Delta i_1(t_1) \Delta i_2(t_2) \rangle}{\bar{i}_1 \bar{i}_2}. \quad (3b)$$

For the generally nonperiodic signals $i_1(t)$ and $i_2(t)$ the angle bracket signifies an average over a time interval long compared with the coherence time t_c

of the light beam (where t_c is the reciprocal of the bandwidth in analogy to (1.2a)). In essence, HBT demonstrated that the fluctuation term (second term of the second equality in (3b)) is greater than zero.

If the incident light beams are sufficiently weak, then, rather than multiplying the fluctuations in the output currents of the two detectors, one could in principle count the arriving photons individually and establish the number of arrival coincidences as a function of detector separation. The nature of the controversial *quantum* effects revealed by this version of intensity interferometry can be made clearer by considering first an experimental configuration in which photons incident upon a *single* detector are counted repeatedly within a prescribed time interval T. (Afterward, we will re-examine intensity interferometry as a split-beam experiment where photons are incident upon two correlated detectors.) The mean number of photoelectrons $\bar{n}$ in the detector output is proportional to the mean light intensity and T. Now if the incident photons, all of which are presumed to have the same polarization, could be thought of as randomly arriving particles, then the variance in count rate would be predicted to be

$$\overline{(\Delta n)^2} \equiv \overline{(n - \bar{n})^2} = \overline{n^2} - \bar{n}^2 = \bar{n}, \tag{4a}$$

where the final equality follows from Poisson statistics. A correct theoretical analysis [7] leads, however, to a variance larger by an amount proportional to the mean number of photons counted

$$\overline{(\Delta n)^2} = \bar{n}\left(1 + \bar{n}\frac{t_c}{T}\right) \tag{4b}$$

under the circumstances (assumed here) that the spectral density of the light is uniform over the optical bandwidth $\Delta v = 1/t_c$. If the light source is unpolarized, then the supplemental term is to be multiplied by one-half, since fluctuations in orthogonal polarizations are independent. What is the origin of this additional spread in count rate?

Purcell, has given a simple, visualizable explanation of relation (4b) in the above-cited reference which cannot be expressed any clearer than in his own words:

If one insists on representing photons by wave packets and demands an explanation in those terms of the extra fluctuation, such an explanation can be given. But I shall have to use language which ought, as a rule, to be used warily. Think, then, of a stream of wave packets, each about $c/\Delta v$ long, in a random sequence. There is a certain probability that two such trains accidentally overlap. When this occurs they interfere and one may find (to speak rather loosely) four photons, or none, or something in between as a result. It is proper to speak of interference in this situation because the conditions of the experiment are just such as will ensure that these photons are in the same quantum state. To such interference one may ascribe the "abnormal" density fluctuations in any assemblage of bosons.

We will see later the wisdom of Purcell's words and the difficulties to which a too cavalier use of the imagery of wave packets can lead. In any event, the

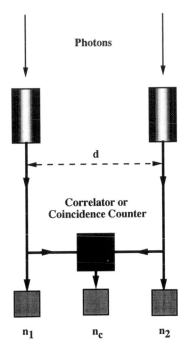

Figure 3.2. Intensity interferometer based on photon counting. The number of photons received at each detector is individually determined (n_1, n_2) as well as the number (n_c) that arrive coincidentally within a specified time window.

broader variance reflects a quantum interference effect deriving from the statistical properties of light as a system of Bose–Einstein particles.

Return to an intensity interferometer configuration (Figure 3.2) where the counts—or more precisely, the fluctuation in counts—at two detectors are correlated. Let n_1 photons be received at detector D_1 and n_2 photons at detector D_2 within a time interval T. For simplicity all photons will be assumed to have identical polarization. The variance in counts at each detector takes the form of relation (4b)

$$\overline{(\Delta n_k)^2} = \bar{n}_k \left(1 + \bar{n}_k \frac{t_c}{T} \right) \qquad (k = 1, 2). \tag{5a}$$

By correlating the two photodetector outputs—i.e., by linking the two outputs together—one has effectively a single-detector configuration again, but with total count rate $n = n_1 + n_2$ and variance

$$\overline{(\Delta n)^2} = \overline{(\Delta n_1 + \Delta n_2)^2}. \tag{5b}$$

Expansion of the right-hand side of expression (5b) and comparison with

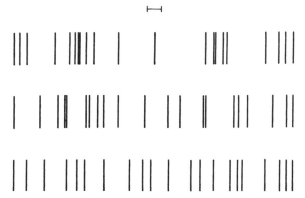

Figure 3.3. Distribution in time of particle counts illustrating bunched (top), random (middle) and antibunched (bottom) arrivals. The horizontal line represents the longitudinal coherence time t_c. (Adapted from Loudon [9].)

relation (4b) leads to the positive cross-correlation

$$\overline{\Delta n_1 \, \Delta n_2} = \bar{n}_1 \bar{n}_2 \frac{t_c}{T}, \tag{5c}$$

observed by HBT. (For an unpolarized light source multiply the right-hand side of (5c) by $\frac{1}{2}$.)

From the standpoint of quantum physics, the nonvanishing correlation between the two components of a split light beam is a consequence of, in Purcell's words, the "clumping" of the photons. The contemporary term, "photon bunching," refers to the tendency of a sequence of photon arrival times (registered at a single photodetector) to be more narrowly spaced than that predicted on the basis of Poisson statistics for randomly occurring events (e.g., the arrival of raindrops), as illustrated in Figure 3.3. The same effect shows up in the number of joint photon arrivals at two detectors, such as in the configuration of Figure 3.2, with the correlator replaced by a coincidence counter. In the case of polarized light coherent over both detecting surfaces, the number of coincident detections n_c within a time window T much longer than the longitudinal coherence time t_c takes the form

$$n_c = n_c^0 \left(1 + \frac{t_c}{T}\right), \tag{5d}$$

where the coincidence count n_c^0 for two uncorrelated beams of light is proportional to the number of counts $(n_1 n_2)$ received at each detector. The supplementary positive term in relation (5d) represents photon bunching, and is seen to have the same form (to within a proportionality factor) as the cross-correlation (5c). It is worth stressing, however, that—as demonstrated by Hanbury Brown's classical wave argument summarized above—the explanation of this phenomenon does not require the introduction of photons.

Indeed, Einstein had examined the same problem, albeit in a different form, in a 1909 summary of the "present state of the problem of radiation" [8], and it is instructive to look at his results. Concerned with the properties of black-body radiation, Einstein showed that the variance in dE_v—the energy of thermal radiation in the spectral range between v and $v + dv$ and volume V—can be written as the sum of two terms as follows:

$$\overline{(\Delta\, dE_v)^2} = hv\, \overline{dE_v} + \frac{c^3}{8\pi v^2 V\, dv}\, \overline{(dE_v)^2}.$$ (6)

The first term corresponds to "shot noise," the fluctuations in energy resulting from fluctuations in the number of particles, each of energy hv. This was the "exciting" part in Einstein's day, for it reflected the grainy or particulate nature of the radiation field and was inexplicable on the basis of classical electromagnetic theory. The first term supported the notion of light quanta and indicated that these quanta (the photons) were subject to the same statistical laws as were the molecules of an ideal gas. On the other hand, the second term—referred to as "wave noise" by HBT—is a purely classical term representing energy fluctuations deriving from fluctuations in the amplitudes of interfering waves. From the standpoint of optical intensity interferometry the quantum shot noise at one detector is uncorrelated with that at the other, and so their product averages away leaving only the correlations due to classical wave noise.

From the foregoing remarks, which pertain largely to the "chaotic" sources [9] employed by HBT—namely, thermal light (e.g., star light) or light from arc lamps—one should not infer that quantum effects play no intrinsic role in the correlations of all light fields. Quite the contrary. States of light, such as so-called "squeezed" light, have been predicted and observed that give rise to correlation effects totally inexplicable within the framework of classical wave theory [10]. The study of these effects falls within the province of quantum optics.

Although still to be worked out more thoroughly and tested experimentally, the quantum optics of *particles* is a conceivable counterpart to that of light, yet embracing processes for which no light-optical analogue actually exists [11]. Thermal electrons for example, like thermal photons, may also give the superficial impression of being randomly emitted, yet arrive correlated in space and time at one or more detectors. There are marked differences, however, in the clustering behavior of thermal electrons and photons owing to differences in particle-number conservation, the tensorial character (vector versus spinor) of the basic fields, and the applicable quantum statistics [12]. The correlations of electrons, or more generally any fermionic system, in contrast to those of photons, requires a quantum description at the very outset, for with only one fermion per quantum state, as required by the Pauli exclusion principle, there is simply no classical wave theory of fermions.

The quantum interference processes, discussed in the first and second chapters, involved exclusively the self-interference of single particles. In the following sections we will consider examples of the interferometry of correlated charged particles that combine various features of the basic Young's two-slit experiment, the Aharonov–Bohm (AB) effect, the Einstein–Podolsky–Rosen (EPR) paradox, and the Hanbury Brown–Twiss (HBT) experiments. These novel processes manifest simultaneously three distinct kinds of quantum interference:

(i) interference, dependent upon optical path-length difference, resulting from the wave-like propagation of particles (or, more accurately, the wave-like description of particle ensembles);

(ii) interference, dependent upon confined magnetic flux, resulting from particle charge and spatial topology; and

(iii) interference, dependent upon quantum statistics, resulting from particle indistinguishability under exchange.

Although the self-interference of single particles is a mystery that cannot be explained, but only described mathematically, physicists have lived with it long enough to develop an intuition for what is likely to occur in given circumstances. However, the study of correlated particle interference in vacuum or in the presence of potential fields is still sufficiently unexplored theoretically and untried experimentally that results can be strange even by the familiar standards of quantum mechanics. In the quantum world there is clearly more than one mystery.

3.2. The AB–EPR Effect with Two Solenoids

It has been noted before in Chapter 1 that the AB effect is an intrinsically nonlocal physical phenomenon in that the spatial distribution of particles diffracting around the solenoid is modified by a magnetic field through which the particles never pass. The nonlocal nature of this effect takes on even stranger dimensions when produced in a context reminiscent of the EPR paradox [13, 14].

In all AB experiments performed to date, the interference pattern is effectively built up by transit of one particle at a time through the apparatus as with a Young's two-slit experiment with weak light source. Consider, however, an experimental configuration like that of Figure 3.4 with two well-separated solenoids and a source that produces pairs of charged particles simultaneously. After production, the two particles separate, each propagating around one of the regions of confined magnetic flux to be

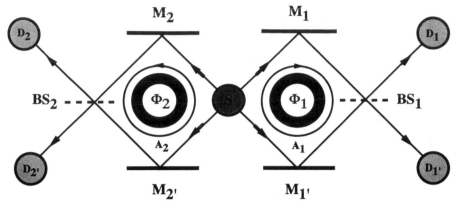

Figure 3.4. Hybrid Aharonov–Bohm and Einstein–Podolsky–Rosen (AB–EPR) experiment. Electrons issue from source S in pairs with one electron entering interferometer 1 and the other electron, whose linear momentum is correlated with the first, entering interferometer 2. The AB effect is manifested, not in the counts at individual detectors, but only in the correlation of counts between one detector at interferometer 1 and another detector at interferometer 2. (Adapted from Silverman [13].)

received afterward at one of two local detectors. The charged particles, which we will suppose to be electrons, are identical and cannot be distinguished, but are required to be correlated in linear momentum—that is, if one particle propagates toward a particular mirror to the right of the source (interferometer 1), the other particle propagates toward a particular mirror to the left of the source (interferometer 2).

To be correlated in this manner, the two electrons had to have once been part of a common system. For example, perhaps the source contains a supply of the exotic atomic species $\mu^+ e^- e^-$ (analogous to the H^- ion). Upon decay of the muon (into a positron, neutrino, and antineutrino) the electrons, which are in a $^1S_0$ ground state, would fly off in opposite directions in order to conserve the total linear momentum of the system. Thus, if one electron propagated toward mirror M_1, the other would propagate toward mirror $M_{2'}$, and likewise for the directions leading to mirrors $M_{1'}$ and M_2. There is nothing intrinsically quantum mechanical about a system at rest splitting into two constituents with the foregoing velocity correlation. It is worth noting, however, that this particular correlation is not required, and that the essential physics would be unchanged if a different momentum correlation were presumed.

We are interested in the following questions: What is the probability of receiving an electron at a left-side detector D_2 or $D_{2'}$ given that an electron was received at one of the right-side detectors, e.g., D_1? And what is the probability of receiving an electron at D_1 *irrespective* of the detector to which the other electron may have gone?

One is tempted to answer these questions according to ideas and imagery drawn from classical physics that the two-electrons—pictured perhaps as mutually receding wave packets of some specified coherence length—propagate independently (except for the velocity correlation posed by momentum conservation) and cease to belong to a common system once the packet overlap has become negligibly small. Then, since the experimental conditions are such that each observer cannot tell by which pathway (source → mirror → detector) a locally detected electron has gone around a solenoid, quantum interference should occur. The electrons, therefore, should give rise independently to AB effects at each end of the double interferometer with a flux-dependent probability of a form similar to that of (1.10d)

$$P_j(\Phi_i) = \alpha_j + \beta_j \cos(\delta_j + 2\pi\Phi_i/\Phi_0), \tag{7a}$$

where $i = 1, 2$, respectively, denotes the right-side or left-side solenoid, and $j = 1, 1', 2, 2'$ designates one of the detectors. Moreover, the joint probability of particle detection—as for any independent events, classical or quantum—would simply be the product of the corresponding probabilities (7a), e.g.,

$$P_{j,k}(\Phi_1, \Phi_2) = P_j(\Phi_1)P_k(\Phi_2). \tag{7b}$$

Reasonable as they may appear, these deductions are not correct.

Let $A(M_1, M_2; D_1, D_2)$ be the amplitude for propagation of one particle via mirror M_1 to detector D_1 and the other particle via mirror M_2 to D_2 with similarly represented amplitudes for the other pathways. Further, let r_1, t_1 be the respective reflection and transmission amplitudes for a particle incident on beam splitter BS_1 from above and r'_1, t'_1 the corresponding amplitudes for incidence from below. The analogous amplitudes for beam splitter BS_2 will be subscripted with 2. It may seem redundant to ascribe left-side and right-side reflection and transmission amplitudes to a device (the beam-splitter) that is apparently symmetric under mirror inversion. Although it is indeed the case that the *probability* of a particle being transmitted or reflected at a lossless beam-splitter does not depend on whether the particle is incident at one side or the other, the corresponding *amplitudes* for these two processes can in fact differ by a relative phase. The origin of this phase may be traced to the mathematical requirement that the transfer matrix (the elements of which are the reflection and transmission amplitudes) connecting input and output states must be unitary. As a consequence of this unitarity, one can derive the relations [15]

$$r_i r'_i - t_i t'_i = |r_i|^2 + |t_i|^2 = 1; \qquad r_i^* t'_i + t_i^* r'_i = 0, \tag{8a}$$

from which it follows that

$$r'_i = r_i^*; \qquad t'_i = -t_i^*. \tag{8b}$$

On the basis of relations (8a, b) one can in the present case take the reflection amplitudes to be purely real ($r_i = r_i' \equiv R_i$) and the transmission amplitudes to be purely imaginary ($t_i = t_i' \equiv iT_i$).

The eight possible amplitudes for joint arrival of particles at the detector pairs (D_1, D_2) or ($D_1, D_{2'}$) are then expressible as

$$A(M_1, M_2; D_1, D_2) = R_1 R_2 \exp[i(\alpha_1 + \alpha_2)], \tag{9a}$$

$$A(M_{1'}, M_{2'}; D_1, D_2) = -T_1 T_2 \exp[i(\alpha_{1'} + \alpha_{2'})], \tag{9b}$$

$$A(M_1, M_{2'}; D_1, D_2) = iR_1 T_2 \exp[i(\alpha_1 + \alpha_{2'})], \tag{9c}$$

$$A(M_{1'}, M_2; D_1, D_2) = iT_1 R_2 \exp[i(\alpha_{1'} + \alpha_2)], \tag{9d}$$

$$A(M_1, M_2; D_1, D_{2'}) = iR_1 T_2 \exp[i(\alpha_1 + \alpha_2)], \tag{9e}$$

$$A(M_{1'}, M_{2'}; D_1, D_{2'}) = iT_1 R_2 \exp[i(\alpha_{1'} + \alpha_{2'})], \tag{9f}$$

$$A(M_1, M_{2'}; D_1, D_{2'}) = R_1 R_2 \exp[i(\alpha_1 + \alpha_{2'})], \tag{9g}$$

$$A(M_{1'}, M_2; D_1, D_{2'}) = -T_1 T_2 \exp[i(\alpha_{1'} + \alpha_2)], \tag{9h}$$

where the relative phases incurred by passage of a particle (with charge e) to one side or the other of the solenoids with vector potential fields $\mathbf{A}_1$ and $\mathbf{A}_2$ take the form of relation (1.6a, b)

$$\alpha_i = (e/\hbar c) \int_{\text{Path } S \to M_i \to BS_k} \mathbf{A}_k \cdot \mathbf{ds}_i \tag{10}$$

with $k = 1$ for $i = 1, 1'$ and $k = 2$ for $i = 2, 2'$. In order not to obscure the principal physics of interest with more complicated notation than is necessary, it has been assumed that:

 (i) the geometrical path lengths of the four specified pathways are equal;
 (ii) the mirrors contribute no differential phase shifts; and
(iii) the detectors have 100% efficiency.

If the momentum correlation of the two particles corresponds to that required by the conservation of linear momentum, i.e., back-to-back separation, then the amplitude for joint detection of one particle at D_1 and the other particle at D_2 (or $D_{2'}$) is given by the linear superpositions

$$A(D_1, D_2) = \frac{1}{\sqrt{2}} [A(M_1, M_{2'}; D_1, D_2) + A(M_{1'}, M_2; D_1, D_2)], \tag{11a}$$

$$A(D_1, D_{2'}) = \frac{1}{\sqrt{2}} [A(M_1, M_{2'}; D_1, D_{2'}) + A(M_{1'}, M_2; D_1, D_{2'})], \tag{11b}$$

The resulting probabilities of joint detection are then

$$P(D_1, D_2) = |A(D_1, D_2)|^2 = \tfrac{1}{2}[(R_1 T_2)^2 + (T_1 R_2)^2 + 2R_1 R_2 T_1 T_2$$
$$\times \cos\{2\pi(\Phi_1 - \Phi_2)/\Phi_0\}], \qquad (12a)$$

$$P(D_1, D_{2'}) = |A(D_1, D_{2'})|^2 = \tfrac{1}{2}[(R_1 R_2)^2 + (T_1 T_2)^2 + 2R_1 R_2 T_1 T_2$$
$$\times \cos\{2\pi(\Phi_1 - \Phi_2)/\Phi_0\}], \qquad (12b)$$

where, by application of Stokes's law,

$$\Phi_1 = \alpha_1 - \alpha_{1'}; \qquad \Phi_2 = \alpha_2 - \alpha_{2'}; \qquad (12c)$$

is the magnetic flux through interferometers 1 and 2, respectively. As indicated by the sense of circulation of the vector potential fields in Figure 3.4, the flux Φ_1 is positive for a magnetic field directed into the paper, and the flux Φ_2 of the other solenoid is positive for a magnetic field directed out of the paper.

The joint detection probabilities of relations (12a, b) manifest a quantum interference that depends on the difference of the magnetic fluxes in solenoids 1 and 2. In other words, from the perspective of one of the observers—let us say the one by detector D_1 on the right side—the number of electrons counted is influenced not only by the nearby solenoid, but *also* by the distant solenoid around which the electrons received at D_1 could never have propagated. If the flux through solenoid 1 is null, i.e., $\Phi_1 = 0$, then the AB effect inferred by the observer at D_1 is attributable *entirely* to the flux Φ_2 in the remote solenoid.

The situation is in fact stranger still. Suppose the arrivals of the electrons emitted into the interferometer on the left side are not observed. Then, from (8a, b), the probability of detecting an electron at D_1 *irrespective* of the path taken by the corresponding electron of the pair is

$$P(D_1) = P(D_1, D_2) + P(D_1, D_{2'}) = \tfrac{1}{2}, \qquad (12d)$$

which is a constant and displays no quantum interference effect at all!

Phrased somewhat differently, an observer counting electrons at D_1 oblivious to the existence of other observers at D_2 and $D_{2'}$—who may be arbitrarily far away—will see no AB effect. Only when he correlates his counts with those of a distant observer is the AB effect manifested. Thus, the inferences drawn by the observer on the right are strongly influenced by the measurements made (or not made) by a remote observer on the left. And yet in either case, as seen from the perspective of the right-side observer, it is the "same" beam of electrons diffracting to one side or the other of his solenoid.

Had the alternative correlation of electrons propagating via mirror pair (M_1, M_2) or $(M_{1'}, M_{2'})$ been chosen, the resulting joint detection probabilities would depend on different combinations of reflection and transmission amplitudes, but the disappearance of the AB effect, as expressed in (12d), would not have changed. The essential point is simply that the electron motion be correlated. If, however, the jointly produced electrons are *uncorrelated*, in which case each pair of pathways from source to mirrors has equal probability, then the resulting joint detection probability, as expected, is simply the product of the single-particle detection probabilities for each detector:

$$P(D_1, D_2) = \tfrac{1}{4}(1 + 2\sqrt{R_1 T_1} \sin(2\pi\Phi_1/\Phi_0))(1 + 2\sqrt{R_2 T_2} \sin(2\pi\Phi_2/\Phi_0)),$$

(13a)

$$P(D_1, D_{2'}) = \tfrac{1}{4}(1 + 2\sqrt{R_1 T_1} \sin(2\pi\Phi_1/\Phi_0))(1 - 2\sqrt{R_2 T_2} \sin(2\pi\Phi_2/\Phi_0)),$$

(13b)

In this case the probability of detecting an electron at D_1 irrespective of the fate of the companion electron now becomes

$$P(D_1) = \tfrac{1}{4}(1 + 2\sqrt{R_1 T_1} \sin(2\pi\Phi_1/\Phi_0)).$$

(13c)

Equation (13c) is a function of the magnetic flux of solenoid 1 only, as would be expected for the ordinary single-solenoid AB effect with uncorrelated single-particle wave packets. The appearance of a sine instead of a cosine as in (7a), merely reflects the 90° phase shift between the amplitudes for reflection and transmission at a beam splitter.

In the foregoing example, the intrinsic spin of the correlated particles has played no role. Let us consider an alternative experimental configuration, shown in Figure 3.5, in which the correlated paths of the particles are directly determined by particle spin. This configuration has the interesting feature of distinguishing fermions and bosons.

We suppose, as before, that the source produces two-particle singlet wave packets, and that the first set of beam splitters, $BS_{1'}$ and $BS_{2'}$, reflect spin-up particles and transmit spin-down particles with 100% probability. This results in correlated paths between the source and detectors involving mirror pair $(M_1, M_{2'})$ for a spin-up particle propagating to the right and spin-down particle propagating to the left, and mirror pair $(M_{1'}, M_2)$ for the opposite situation. The spin-statistics connection, however, gives rise to a relative phase of 0 ($e^{i0} = +1$) or π ($e^{i\pi} = -1$) between the amplitudes for the direct and exchange processes depending on whether the particles are bosons or fermions, respectively. The beam splitters BS_1 and BS_2 are not sensitive to spin and have the reflection and transmission amplitudes R_1, R_2, iT_1, iT_2 as designated previously.

By an analysis largely the same as that of the experimental configuration (Figure 3.4) insensitive to spin, one may demonstrate that the probabilities for joint arrival of particles at the detector pairs (D_1, D_2) and $(D_1, D_{2'})$ are

$$P(D_1, D_2) = \tfrac{1}{2}[(R_1 T_2)^2 + (T_1 R_2)^2 \pm 2R_1 R_2 T_1 T_2 \cos(2\pi(\Phi_1 - \Phi_2)/\Phi_0)],$$ (14a)

$$P(D_1, D_{2'}) = \tfrac{1}{2}[(R_1 R_2)^2 + (T_1 T_2)^2 \mp 2R_1 R_2 T_1 T_2 \cos(2\pi(\Phi_1 - \Phi_2)/\Phi_0)],$$ (14b)

where upper and lower signs of the interference terms, respectively, refer to

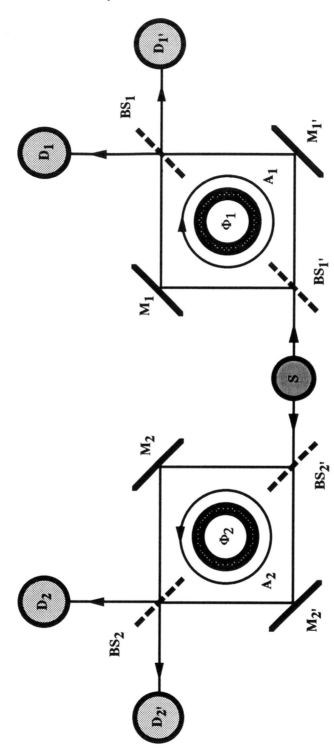

Figure 3.5. Alternative configuration of the AB–EPR experiment sensitive to particle spin. The correlation in particle counts between one detector at interferometer 1 and a second detector at interferometer 2 can reveal whether the particles are fermions or bosons. (Adapted from Silverman [13].)

bosons and fermions. As before, the single-particle detection probability shows no quantum interference, i.e., $P(D_1) = \frac{1}{2}$. Also, if the particles are again uncorrelated in their motions—i.e., if one particle can take any available pathway irrespective of the pathway taken by the other particle—then the detectors on one side of the interferometer would register a quantum interference effect independent of the detectors on the other side, and, of course, the joint detection probability would factor just as in relation (7b).

When looked at with the "rational" expectations drawn from classical physics, there is a sort of double irony in the experimental outcomes of these AB–EPR *Gedankenexperiments*. First, as with the ordinary AB effect, the *local* magnetic field *through* which electrons do not pass can influence the electron spatial distribution. And second, the *distant* magnetic field *around* which electrons do not pass can also influence their spatial distribution. In other words, the act of *not* observing where electrons go on one side of the interferometer apparently destroys the quantum interference of the detected electrons on the other side of the interferometer. There is no restriction in principle on how far apart the separated pair of electrons can be at the time of detection.

Although these results seem to defy "common sense"—which, of course, they do, for common sense is rarely tutored by quantum mechanical experiences—they are a direct consequence of the entanglement of the electron wave function as represented, for example, by expressions (11a, b). In general, the wave function of a multiparticle system is entangled if it cannot be factored into a product of single-particle wave functions. From the standpoint of quantum mechanics the entangled system of particles remains a *single* system despite subsequent separation of its components.

The foregoing examples illustrate in a novel context, as yet unrealized in the laboratory, the "ghostly" long-range correlations intrinsic to quantum mechanics that physicists have been struggling to understand for over half a century.

3.3. The AB–HBT Effect in a Two-Slit Interferometer

In the first chapter of this book we introduced the AB effect by means of a Young's two-slit experiment with a flux-bearing solenoid placed between the slits. Let us now reconsider this experimental configuration modified, as shown in Figure 3.6, by the use of a source which generates pairs of electrons and the addition of a second detector at the viewing screen [16]. Of interest is not only the probability of electron arrival at each detector singly, but also, as in the example of Section 3.2, the joint probability of electron arrival at the two detectors. For this purpose, the outputs of the two detectors in Figure 3.6 are schematically linked to a correlator

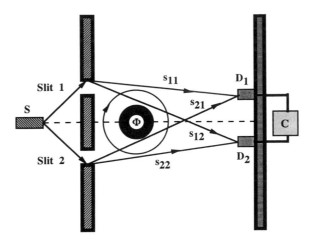

Figure 3.6. Hybrid AB–HBT experiment. Pairs of electrons issuing from source S diffract around the solenoid and are received at two detectors whose outputs are correlated. The electron antibunching revealed by the correlated counts is unaffected by the magnetic flux although the count rate of singly emitted electrons manifests the AB effect. (Adapted from Silverman [16].)

characteristic of the HBT experiment. In particular, we would like to ascertain the effect of the confined magnetic field on the probability of coincident electron detection.

One's intuition, based on the AB and AB–EPR effects and a general "feel" for quantum mechanics, would likely suggest that the joint detection probability will be some harmonic function of the confined magnetic flux. But quantum mechanical intuition can be just as misleading as classical mechanical common sense, and the actual theoretical outcome is perhaps even stranger than what was first reported in the literature [17]. If the AB effect is a subtle one, then the two-slit AB–HBT experiment, which involves the quantum states of identical particles, is even more so and illustrates strikingly the potential pitfalls of adopting too literally the visual imagery of wave packets.

We assume again for simplicity—although the assumption can be readily relaxed in a more general analysis—that the electrons produced by the source are spin-polarized so they can interfere, and that one electron is to issue from each slit and be received at each detector. (By design of the apparatus, the passage of two particles through one slit and none through the other can be made negligibly small. Also, the arrival of two particles at one detector and none at the other would give rise to a null coincidence count, and therefore contribute nothing to the signal.)

In contrast to the momentum correlation required of the (spin-independent) AB–EPR effect, the entanglement of the electrons in the present case arises in an entirely different way. From a classical perspective, one can imagine two alternative emission-detection processes:

(i) the arrival at detectors D_1 and D_2 of electrons that issued, respectively, from slits 1 and 2; and

(ii) the converse situation with the particles exchanged.

Because it is impossible to distinguish these two alternatives experimentally (without an intrusive observation that would disturb the system and destroy the quantum interference), the correct quantum mechanical procedure requires that the wave function characterizing the arrival of the electrons at the two detectors be antisymmetrized under particle exchange as follows:

$$\Psi(1, 2) = \frac{1}{\sqrt{2}} [\phi_1(D_1)\phi_2(D_2) - \phi_1(D_2)\phi_2(D_1)], \tag{15a}$$

where $\phi_i(D_j)$ represents the amplitude of the electron that has propagated from the ith slit to the jth detector. In the absence of the solenoid, this amplitude could be appropriately represented by a wave packet of the form

$$\phi_i(D_j) = \int g(k) \exp(iks_{ij}) \, dk \equiv \varphi(s_{ij}), \tag{15b}$$

comprising a linear superposition of plane wave states of wave number k and amplitude $g(k)$ that have traversed a geometric path length s_{ij} between slit and detector. In the presence of the vector potential field of the solenoid, however, each single-particle amplitude is the product of two factors, one of geometric and the other of magnetic origin

$$\phi_i(D_j) = \varphi(s_{ij}) \exp(i\alpha_{ij}), \tag{15c}$$

where, as in (10) and (12c), the phase shifts engendered by the field of the solenoid

$$\alpha_{ij} = (e/\hbar c) \int_{\text{Source–Slit } i\text{–Detector } j} \mathbf{A} \cdot \mathbf{ds}, \tag{15d}$$

are again related to the confined magnetic flux Φ by Stokes's law

$$a_{11} - \alpha_{21} = \alpha_{12} - \alpha_{22} = (e/\hbar c)\Phi = 2\pi\Phi/\Phi_0. \tag{15e}$$

The joint probability that one electron is received at each detector (to which the coincidence count rate is proportional) is

$$P(D1, D2) = |\Psi(D1, D2)|^2, \tag{16a}$$

which, upon substituting relations (15b, c, d) into (15a), leads to an expression of the form

$$\begin{aligned} P(D_1, D_2) = (\tfrac{1}{2})[|\varphi(s_{11})\varphi(s_{22})|^2 + |\varphi(s_{12})\varphi(s_{21})|^2 \\ - 2 \, \text{Re}\{\varphi(s_{11})\varphi(s_{22})\varphi(s_{12})^*\varphi(s_{21})^*\}]. \end{aligned} \tag{16b}$$

There is indeed a quantum interference term, but all dependence on the magnetic flux has vanished! The magnetic phase shifts for the direct and exchange processes have mutually canceled.

Suppose one of the detectors (e.g., D_2) is turned off so that only electrons at the other detector (D_1) are counted. What the inactive detector now receives is seemingly irrelevant, and it may therefore appear reasonable that the total electron amplitude at the active detector is the linear superposition of electron "waves" from slits 1 and 2, i.e.,

$$\Psi(D_1) \sim \varphi(s_{11}) \exp(i\alpha_{11}) + \varphi(s_{21}) \exp(i\alpha_{21}), \tag{17a}$$

from which it would follow that the (appropriately normalized) single-particle detection probability

$$P(D_1) = |\Psi(D_1)|^2$$
$$= (\tfrac{1}{2})[|\varphi(s_{11})|^2 + |\varphi(s_{21})|^2 + 2\,\mathrm{Re}\{\varphi(s_{11})\varphi(s_{21})^* \exp(i2\pi\Phi/\Phi_0)\}] \tag{17b}$$

clearly depends on the confined magnetic flux. One must be careful, however. Can it truly be the case that the AB effect occurs if one looks for electrons at *one* location, and does not occur if one looks for electrons at *two* locations? If so, the present example leads to an extraordinarily puzzling consequence which I illustrate concretely by resorting to a monochromatic plane-wave description of the electrons. The above expressions for single and joint detection probabilities then reduce to the following relations for electrons of wave number k:

$$P(D_1) = (\tfrac{1}{2})[1 + \cos\{k(s_{11} - s_{21}) + 2\pi\Phi/\Phi_0\}], \tag{18a}$$

$$P(D_2) = (\tfrac{1}{2})[1 + \cos\{k(s_{12} - s_{22}) + 2\pi\Phi/\Phi_0\}], \tag{18b}$$

$$P(D_1, D_2) = (\tfrac{1}{2})[1 + \cos\{k(s_{11} - s_{21} + s_{22} - s_{12})\}]. \tag{18c}$$

By arranging experimental conditions so that the geometrical phases are

$$k(s_{11} - s_{21}) = k(s_{22} - s_{12}) = -\pi/2,$$

(i.e., symmetrical disposition of D_1 and D_2 above and below the forward beam direction) and adjusting the magnetic flux so that $\Phi/\Phi_0 = \tfrac{1}{4}$, one deduces that $P(D_1, D_2) = 1$, $P(D_1) = 1$, and $P(D_2) = 0$. How can it be that there is a 100% coincidence count rate if the individual count rate at one of the detectors is zero?

It is the inconsistent treatment of correlated and uncorrelated electron states that lies at the origin of the foregoing paradoxical results. The probabilities $P(D_1)$ and $P(D_2)$ are derived from *single-particle* relations (17a, b) and therefore depict the case of *uncorrelated* electron propagation through the two slits. Although it may seem reasonable, when only one of the two detectors is registering particles, to imagine the electrons as arriving independently at the detector in single-particle wave packets, this is not correct. The joint probability $P(D_1, D_2)$ for uncorrelated particles is simply the product $P(D_1)P(D_2)$ and is, as expected, a function of the magnetic flux.

For correlated electron pairs, however, the probability that one particle arrives at a particular detector (D_1) irrespective of the subsequent fate of the unobserved companion electron is determined by integrating the joint probability over the full coordinate range of the second detector. Thus, for the special case of plane-wave states, one can show that relation (18c) leads to

$$\bar{P}(D_1) = \int_{-\infty}^{+\infty} P(D_1, D_2)\, dD_2 \sim (\tfrac{1}{2})[1 - J_0(kd)\cos(kd\sin\theta)], \quad (18d)$$

where $J_0(kd)$ is the *zeroth* order Bessel function, d is the slit separation, and θ is the angle (with respect to the forward direction) of detector D_1 seen from the midpoint between the slits. (As in the case of ordinary Fraunhofer diffraction, the approximate far-field relation $s_{21} - s_{11} \sim d\sin\theta$ has been assumed.) Relation (18d) does not depend on magnetic flux and cannot vanish for nonvanishing $P(D_1, D_2)$. Whether or not an actual coincidence experiment of the type proposed manifests a flux dependence therefore depends on the quantum composition of the electron beam, i.e., on the nature of the particle correlations. This point will be discussed further in the next section.

Although emphasis has so far been placed on the effect (or noneffect!) of the isolated magnetic field, the physical implication of the interference term displayed in (16b) or (18c) is also of interest, for it is indicative of a type of electron avoidance behavior termed "antibunching." The joint probability of coincident electron detection vanishes when D_1 is coincident with D_2, i.e., $s_{11} - s_{21} = s_{12} - s_{22}$. In contrast to the "bunching" of photons manifested in the HBT experiments, indistinguishable electrons tend never to arrive in pairs.

3.4. Correlated Particles in a Mach–Zehnder Interferometer

The Young's two-slit configuration represents what in optics is termed a wave-front splitting interferometer. Portions of a primary incident wave front give rise either directly or indirectly to coherent sources of secondary waves. An alternative type of interferometer, of which the Mach–Zehnder type schematically shown in Figure 3.7 is an example, makes use of amplitude splitting, i.e., the division of an incident beam into two components which travel different paths before recombining and interfering. Although the wave-front splitting configuration of the previous section resulted in a quantum cancellation effect for the two-particle AB–HBT experiment, this need not be the case for other interferometer configurations. In this section we will examine the interferometry of correlated fermions in a

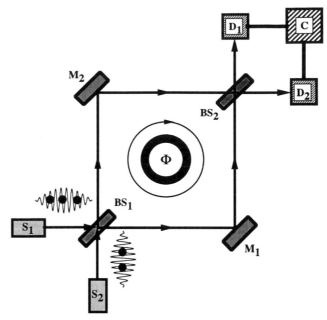

Figure 3.7. AB–HBT experiment employing a four-port Mach–Zehnder interferometer. Electrons from one or two sources enter the interferometer and propagate around a region of confined magnetic flux. The cross-correlation in counts at the two detectors manifests an electron antibunching sensitive to the magnetic flux Φ. (Adapted from Silverman [18].)

Mach–Zehnder interferometer paying particular attention to the nature of possible fermion correlations and the effects of external potential fields on fermion fluctuations [18].

It is well known theoretically and verified experimentally that photons (which are massless bosons) can manifest a variety of "clustering" behaviors depending upon the nature of the source. For example, it has already been mentioned that the fluctuations of chaotic light such as black-body radiation manifest a positive cross correlation, or bunching. However, laser light above threshold, characterized by so-called coherent states [19], ideally gives rise to no fluctuations. The coherent state—which represents, as closely as quantum mechanics allows, the motion of a classical harmonic oscillator—has the same coherence properties as a classical stable wave of well-defined amplitude and phase. Finally, under appropriate circumstances the fluorescent emission from single two-level atoms manifests an anti-bunching effect as characterized by a correlation function, (3b), less than unity [20]. The photon antibunching arises from the fact that the atom can radiate only from its excited state. Thus, having emitted one photon, it cannot emit a second until it has been re-excited.

What about electrons? Is it the case, as has long been thought, that the antisymmetrization of multielectron wave functions required by the Pauli principle *always* leads to electron *anti*correlation—i.e., fluctuations in which electrons tend to avoid arriving in pairs? Surprisingly (perhaps), the answer is "no" [21].

In the general case, particles distributed in some specifiable way over linear momentum and spin states can enter the interferometer of Figure 3.7 through either or both of the two entrance ports. The entering beam is divided at beam splitter BS_1, reflects from mirrors M_1 and M_2 to propagate around a tube of magnetic flux confined to the center of the interferometer. At beam splitter BS_2 it is divided again and directed into detectors D_1 and D_2 whose outputs are correlated as in the HBT experiment. It is assumed that the mirrors are 100% reflecting and that no particle loss occurs at the beam splitters whose reflectance and transmittance amplitudes are, respectively, (r_1, t_1) and (r_2, t_2) for beams incident at the left surface and (r_1', t_1') and (r_2', t_2') for beams incident at the right. The relations between these amplitudes have been given by (8a, b).

As in the case of photon interference with multiphoton states where discrete attributes of light play an important role, the analysis of the Mach–Zehnder fermion interferometer is also most easily effected by means of a field-theoretic (or "second-quantized") description [22]. In this description the fermion fields—which superficially resemble the wave functions of familiar "first-quantized" quantum mechanics—are not functions, but operators independent of the state of the system; all information concerning the latter is represented by the density or statistical operator ρ. The fermion field operators at the two input and output ports of the interferometer of Figure 3.7 can be expressed in a plane-wave basis as follows (with $i = 1, 2$):

$$\Psi_i^{(in)}(x, t) = \sum_{k, s} b_i(k, s) \exp[i(kx - \omega t)], \tag{19a}$$

$$\Psi_i^{(out)}(x, t) = \sum_{k, s} d_i(k, s) \exp[i(kx - \omega t)], \tag{19b}$$

where the annihilation $[b_i(k, s)$ and $d_i(k. s)]$ and creation $[b_i(k. s)^\dagger$ and $d_i(k, s)^\dagger]$ operators satisfy the standard fermion anticommutation relations of the form

$$\{b_i(k, s), b_j(k', s')\} \equiv b_i(k, s)b_j(k', s') + b_j(k', s')b_i(k, s) = \delta_{ij}\delta_{ss'}\delta(k - k'). \tag{20}$$

By taking account of the changes in phase and amplitude resulting from free propagation between entrance and exit ports and interaction at each beam splitter one can establish to within a global phase factor the following relationship between input and output annihilation operators;

$$\begin{pmatrix} d_1 \\ d_2 \end{pmatrix} = \begin{pmatrix} t_1 t_2' e^{-i\theta/2} + r_1 r_2 e^{i\theta/2} & r_1' t_2' e^{-i\theta/2} + t_1' r_2 e^{i\theta/2} \\ t_1 r_2' e^{-i\theta/2} + r_1 t_2 e^{i\theta/2} & r_1' r_2' e^{-i\theta/2} + t_1' t_2 e^{i\theta/2} \end{pmatrix} \begin{pmatrix} b_1 \\ b_2 \end{pmatrix}, \tag{21a}$$

where θ is the total phase difference incurred between the "upper" (via M_2) and "lower" (via M_1) beam components. Upon implementation of the unitarity relations (8a, b), equation (21a) leads to

$$d_1(k, s) = (|r_1 r_2| e^{i\theta/2} + |t_1 t_2| e^{-i\theta/2}) b_1(k, s) + i(|r_1 t_2| e^{-i\theta/2} - |t_1 r_2| e^{i\theta/2}) b_2(k, s),$$
(21b)

$$d_2(k, s) = (-i|t_1 r_2| e^{-i\theta/2} - |r_1 t_2| e^{i\theta/2}) b_1(k, s) + (|r_1 r_2| e^{-i\theta/2} + |t_1 t_2| e^{i\theta/2}) b_2(k, s).$$
(21c)

It will be assumed in what follows that the beam is quasi-monochromatic, so that the momentum spread is much less than the mean particle momentum ($\Delta k \ll k_0$), in which case θ is independent of the momentum distribution to a good approximation. If the relative phase is produced entirely by the AB effect, then $\theta = 2\pi\Phi/\Phi_0$ depends only on an external parameter (the magnetic flux) and is rigorously independent of the geometric path length and momentum distribution of the particles. Since we are concerned principally with the quantum effects of fields and statistics, we adopt at the outset a square interferometer geometry to eliminate the relative phase arising from unequal geometric path lengths. This simplification can be readily relaxed whenever necessary, and the resulting relative phase easily determined.

From the fermion creation and annihilation operators one can construct (as in the standard quantum mechanical analysis of an harmonic oscillator) Hermitian operators corresponding to the number of particles entering the input ports ($i = 1, 2$)

$$N_i = \sum_{k, s} b_i(k, s)^\dagger b_i(k, s),$$
(22a)

and the number of particles leaving the output ports to be received, respectively, at detectors D_1 and D_2

$$N(D_i) = \sum_{k, s} d_i(k, s)^\dagger d_i(k, s).$$
(22b)

Substitution of relations (21b, c) into (22b) for the special, but useful, case of 50–50 beam splitters ($|r_i| = |t_i| = 0.5$) allows one to relate the input and output particle-number operators as follows:

$$N(D_1) = \tfrac{1}{2}[N_1 \cos^2(\theta/2) + N_2 \sin^2(\theta/2) + M \cos(\theta/2) \sin(\theta/2)], \quad (23a)$$

$$N(D_2) = \tfrac{1}{2}[N_1 \sin^2(\theta/2) + N_2 \cos^2(\theta/2) - M \cos(\theta/2) \sin(\theta/2)]. \quad (23b)$$

In the preceding relations

$$M = \sum_{k, s} [b_1(k, s)^\dagger b_2(k, s) + b_2(k, s)^\dagger b_1(k, s)]$$
(23c)

is an operator that removes a particle from the beam at one input port and adds it to the beam of the other input port. Note that

$$N(D_1) + N(D_2) = N_1 + N_2 \equiv N, \tag{23d}$$

as is required by conservation of particle number. For charged particles, this is tantamount to the conservation of electric charge, as well.

It is worth noting that if the second beam splitter is opaque ($|r_2| = 1$; $|t_2| = 0$) the configuration is equivalent to an electron version of the original HBT split-beam photon-counting experiment. The resulting operator expressions are similar to those of (23a, b) but without the trignonometric factors, since the configuration does not constitute an interferometer, and there is consequently no phase angle.

The characteristics of any input beam are specified by the mathematical form of the density operator ρ; theoretical expressions corresponding to the expectation values of dynamical observables are determined by taking the trace of the appropriate operator with ρ. Thus, the mean number of counts received by each detector in a given sampling time (the time during which a known number of particles enters the interferometer) is

$$\overline{N(D_i)} = \text{Tr}\{\rho N(D_i)\}, \tag{24a}$$

the number of coincident counts at the two detectors is

$$\overline{N(D_1)N(D_2)} = \text{Tr}\{\rho N(D_1)N(D_2)\}, \tag{24b}$$

and the cross-correlation in fluctuations in counts at the two detectors is

$$C(D_1, D_2) = \text{Tr}\{\rho(N(D_1) - \overline{N(D_1)})(N(D_2) - \overline{N(D_2)})\}$$
$$= \overline{N(D_1)N(D_2)} - \overline{N(D_1)}\,\overline{N(D_2)}. \tag{24c}$$

Consider the simple case of a beam of electrons of reasonably well-defined linear momentum and energy entering the interferometer through only one of the two entrance ports, e.g., port 1. In a basis $|n_1\{k, s\}; n_2\{k, s\}\rangle$ of eigenstates of the input particle number operators, relation (22a), the density operator of the proposed system would consist of states of the form $|n_1\{k, s\}; 0\rangle$ where the designation $n_i\{k, s\}$ is the total number of particles entering port i ($i = 1, 2$) with a given spectrum of linear momentum and spin eigenvalues. (Note that the eigenvalues of N_i are independent of how the particles are distributed over momentum and spin states.) In the case characteristic of thermal and field-emission electron sources, where the number of particles in the input beam fluctuates about a mean value $\overline{N_i}$ with dispersion $(\Delta N_1)^2$, the density operator is constructed from a mixture of states of different particle number

$$\rho_{\text{chaotic}} = \sum_{\{k, s, n\}} \rho(n_1\{k, s\})|n_1\{k, s\}; 0\rangle\langle n_1\{k, s\}; 0|. \tag{25a}$$

The source is designated chaotic in analogy to chaotic optical sources for which the density operator is diagonal in a photon number basis.

Substitution of density matrix (25a) into the expressions (24a–c) leads to mean particle counts at each detector

$$\overline{N(D_1)} = (\overline{N_1}/2)[1 + \cos\theta], \tag{25b}$$

$$\overline{N(D_2)} = (\overline{N_1}/2)[1 - \cos\theta], \tag{25c}$$

the joint count rate

$$\overline{N(D_1)N(D_2)} = \tfrac{1}{4}\overline{N_1}(\overline{N_1} - 1)\sin^2\theta, \tag{25d}$$

and the cross-correlation in particle fluctuations

$$C(D_1, D_2) = \tfrac{1}{4}[(\Delta N_1)^2 - \overline{N_1}]\sin^2\theta, \tag{25e}$$

all of which clearly show an influence of the magnetic flux. Note that the coincidence count rate, (25d), vanishes, as it must, for an input of $\overline{N_1} = N_1 = 1$, since the one particle must be received at only one of the detectors, and thus there can be no coincidence. This result contrasts with that deduced from a classical wave analysis in which a split incident beam of arbitrarily weak amplitude could still illuminate both detectors.

For a beam of randomly emitted classical particles the fluctuation in particle number is governed by Poisson statistics for which the variance is equal to the mean, and the cross-correlation (25e) vanishes. However, for a chaotic electron source, the antisymmetrization of the state vectors leads to a variance that is smaller than the mean incident particle number by an amount proportional to what is termed the beam degeneracy δ. In the experimentally realistic case of a counting interval T long in comparison to the beam coherence time t_c, the variance takes the form [23]

$$(\Delta N_1)^2 = \overline{N_1}\left[1 - \overline{N_1}\sigma F(0)\frac{t_c}{T}\right], \tag{26a}$$

where σ is a polarization factor (unity for total spin polarization and one-half for an unpolarized beam), and $F(0)$ is a coherence function (unity for a beam cross section equal to a coherence area, i.e., an area on the order of the square of the transverse coherence length l_t). Expression (26a) is the fermion analogue of (4b) for thermal light. The beam degeneracy, determined practically from

$$\delta = \overline{N_1}\frac{t_c}{T}, \tag{26b}$$

is interpretable as the mean number of particles per cell of phase space. This number can never exceed unity for fermions, since there can be at most one fermion per quantum state. Substitution of (26a, b) into relation (25e) leads to a negative cross-correlation

$$C(D_1, D_2) = -\tfrac{1}{4}\overline{N_1}\sigma F(0)\delta\sin^2\theta, \tag{26c}$$

indicative again of electron antibunching.

Although a vector potential field is ostensibly responsible for the quantum phase θ in the foregoing analysis, one can envision different experimental configurations for which the phase is attributable to other external potentials. One interesting example is that of gravity [24]. The force of gravity is the dominant influence in shaping the macroscopic universe, but it is intrinsically so weak that its effects on the elementary constituents of matter are for the most part negligible. There are very few physical systems—in particular, systems directly accessible to laboratory investigation—whose dynamical behavior requires a quantum mechanical description with simultaneous inclusion of a gravitational interaction.

One example, however, is the Colella–Overhauser–Werner (COW) experiment [25] which provided the first experimental demonstration of the effect of gravity in a context unique to quantum mechanics, i.e., through a gravitational influence on the phase of a wave function leading to quantum interference. The COW experiment employed a neutron beam split and recombined in a single-crystal neutron interferometer, a configuration effectively equivalent to that shown in Figure 3.7 but with the solenoid removed and the plane of the interferometer oriented vertically so that the "upper" path (between mirror M_2 and beam splitter BS_2) is actually a distance z above the "lower" path (between BS_1 and M_1). Then, for a nonrelativistic particle of mass m and mean momentum $\hbar k_0$ there is a gravitationally induced phase shift

$$\theta = m^2 gzd/\hbar^2 k_0, \qquad (27)$$

where g is the local acceleration of gravity and d is the horizontal path length between a mirror and beam splitter at the same height. Equations (25b, c) illustrate the effect of the gravitationally induced quantum interference on the particle distribution at the output ports. Similarly, (25e) or (26c) manifests the effect of gravity on particle correlations. Although the gravitational field is present throughout the interferometer and the particles are therefore subject to the force of gravity, (in contrast to the AB effect where the magnetic field is confined to the interior of the solenoid and the particles are not subject to a magnetic force), it is *not* the gravitational force, but rather the gravitational potential, that plays a direct role here. The relative phase expressed in (27), which is proportional to the area (zd) of the interferometer, may be thought to accrue only over the horizontal segments of the motion where the classical effect of the gravitational force on the split beam is zero, since no work is done.

The utilization of both input ports of a fermion interferometer permits, at least in principle, the construction of fermion ensembles manifesting a variety of new and surprising statistical properties. One finds that the clustering behavior of fermions need not be limited to antibunching by Fermi–Dirac statistics, but, like light, can depend as well on the specific state composition of the ensemble. To be sure, each multiparticle basis state that contributes to the ensemble description must, itself, be compatible with the antisymmetrization restrictions of quantum statistics. Nevertheless, there is room for considerable diversity among possible fermionic ensembles.

Of particular interest are fermion beams described by basis states that are labeled by the total number of particles entering the two ports and by the difference in number of particles between the two input ports. For such states one does not in general know the number of particles entering each port individually. It will be seen that these states span representations of an SU(2) algebra constructed from the fermion creation and annihilation operators. In orther words, the mathematical description is in many ways similar to that of the familiar treatment of orbital or spin angular momentum. There is a vector operator $\mathbf{J}$ with components J_i ($i = x, y, z$) from which can be constructed "raising" and "lowering" operators J_+ and J_- that effectively transfer particles between input ports, and a Casimir operator $J^2 = \mathbf{J} \cdot \mathbf{J}$ which, under appropriate circumstances, gives the total number of particles entering the interferometer through the two ports. (A Casimir operator is a nonlinear invariant operator that commutes with all other members of the algebra; in general, there are $n - 1$ of them for the special unitary group SU(n).) The basis states $|j, m\rangle$, which will be termed correlated two-port states, are labeled by the quantum numbers specifying the eigenvalues of J^2 ($= j(j + 1)$) and J_z, in contrast to the uncorrelated two-port states $|n_1; n_2\rangle$ which are eigenvectors of the input number operators N_1 and N_2. To aid in avoiding confusion between these two different representations, the quantum numbers of uncorrelated two-port basis states will be separated by a semicolon (;) while those of correlated two-port basis states will be separated by a comma (,).

Let us now look at the development of the correlated two-port states in detail [21, 26]. From the fermion annihilation and creation operators one can construct the following bilinear superpositions:

$$J_x = \tfrac{1}{2} \sum_{k, s} [b_1(k, s)^\dagger b_2(k, s) + b_2(k, s)^\dagger b_1(k, s)], \tag{28a}$$

$$J_y = \frac{1}{2i} \sum_{k, s} [b_1(k, s)^\dagger b_2(k, s) + b_2(k, s)^\dagger b_1(k, s)], \tag{28b}$$

$$J_z = \tfrac{1}{2} \sum_{k, s} [b_1(k, s)^\dagger b_1(k, s) + b_2(k, s)^\dagger b_2(k, s)] = \tfrac{1}{2}(N_1 - N_2), \tag{28c}$$

and the Casimir invariant

$$J^2 = \frac{N}{2}\left(\frac{N}{2} + 1\right) - W, \tag{28d}$$

where N is the total particle number operator in (23d), and

$$W = \sum_{k, s; k', s'} [b_1(k, s)^\dagger b_2(k', s')^\dagger b_2(k, s) b_1(k', s')$$
$$+ b_1(k, s)^\dagger b_2(k', s')^\dagger b_2(k', s') b_1(k, s)] \tag{28e}$$

is an operator that transfers two particles between input ports. The vector operator $\mathbf{J}$ commutes with both N and W, although N and W do not commute with each other.

When the relative phase θ incurred by passage of a wave packet through the interferometer is independent of the particle momentum and spin (as is the case for the AB effect)—or very nearly so in the case of a quasi-monochromatic beam—the particle number operators for both input and output ports are expressible in terms of the above SU(2) operators in a relatively simple way:

$$N_1 = \tfrac{1}{2}N + J_z, \tag{29a}$$

$$N_2 = \tfrac{1}{2}N - J_z, \tag{29b}$$

$$N(D_1) = \tfrac{1}{2}N + J_z \cos\theta + J_x \sin\theta, \tag{29c}$$

$$N(D_2) = \tfrac{1}{2}N - J_z \cos\theta - J_x \sin\theta. \tag{29d}$$

It is also useful to denote the operator $N(D)$ which gives the difference in particle counts at the output ports

$$N(D) = N(D_1) - N(D_2) = 2[J_z \cos\theta + J_x \sin\theta]. \tag{29e}$$

Recall from relation (28c) that $2J_z$ gives the particle number difference at the input ports. In anticipation of our interest in the cross-correlation of particle counts at the two output ports, we record as well the product

$$N(D_1)N(D_2) = \tfrac{1}{2}[N(D_1)^2 + N(D_2)^2 - N(D)^2], \tag{29f}$$

which follows straightforwardly from squaring $N(D)$ in the first equality of (29e).

By means of the raising and lowering operators

$$J_+ = J_x + iJ_y = \sum_{k,s} b_1(k,s)^\dagger b_2(k,s), \tag{30a}$$

$$J_- = J_x - iJ_y = \sum_{k,s} b_2(k,s)^\dagger b_1(k,s), \tag{30b}$$

one can construct all members of a family of states $|j, m\rangle$ of given j by knowing one of the states. For example, start with the state that corresponds to all N particles entering the interferometer through port 1, i.e., $|N; 0\rangle$ in the uncorrelated two-port representation. This state is an eigenstate of W with eigenvalue 0, and consequently from relations (28c, d) an eigenstate of J^2 and J_z with quantum numbers $j = m = N/2$ in the correlated two-port representation. By sequential application of J_- to the correlated two-port state $|N/2, N/2\rangle$, one can generate the full spectrum of states $|j, m\rangle$ in the manner below

$$J_-^{(\frac{N}{2} - m)}\left|\frac{N}{2}, \frac{N}{2}\right\rangle = \left[\frac{N!((N/2) - m)!}{((N/2) + m)!}\right]^{1/2}\left|\frac{N}{2}, m\right\rangle. \tag{31}$$

As an explicit illustration of the connection between the two representations, consider a two-particle input beam $(N = 2)$ where the particles can have momentum-spin values (k_α, s_α) or (k_β, s_β). To avoid encumbering our notation with unnecessary symbols, the set of eigenvalues (k_γ, s_γ) will be represented simply by the label γ. Thus, the state corresponding to two particles entering port 1 and no particles entering port 2 can be expressed by the state vectors

$$|1, 1\rangle = |2; 0\rangle = |\{\alpha, \beta\}; 0\rangle, \tag{32a}$$

where the first vector is a $|j, m\rangle$ state and the second vector is an $|n_1; n_2\rangle$ state specified in greater detail by the third vector which is implicitly antisymmetric under exchange of particle labels α, β. Application of J_- to relation (32a) leads to the state

$$|1, 0\rangle = |1; 1\rangle = -\frac{1}{\sqrt{2}}[|\alpha; \beta\rangle - |\beta; \alpha\rangle], \tag{32b}$$

in which one particle enters each port, but we do not know through which port a particle of particular spin and linear momentum goes. There is a 50% probability for each particle to enter through each port. Finally, application of J_- a second time results in the state

$$|1, -1\rangle = |0; 2\rangle = -|0; \{\alpha, \beta\}\rangle \tag{32c}$$

in which both particles enter port 2 and none enters port 1.

Since the correlated two-port states cannot in general be factored into state vectors of the two ports individually, they exhibit a novel form of entanglement beyond that deriving strictly from the spin-statistics connection. It is also important to note—and easily verified in the special case of the above example—that a system characterized by a density operator diagonal in a representation of SU(2) correlated states is generally *not* diagonal in a basis of momentum-spin states. Thus, the fluctuation behavior characteristic of the correlated states need not necessarily be similar to that of the chaotic states examined previously.

Let us first examine the quantum statistical behavior of an ensemble of entering fermions characterized by a density operator diagonal in a basis of SU(2) correlated states. From (29c–f), the cross-correlation of outputs at D_1 and D_2 can be expressed in the form

$$C(D_1, D_2) = \frac{1}{4}(\Delta N)^2 - (\Delta m)^2 \cos^2 \theta - \frac{1}{2}(\frac{1}{4}\overline{N^2} + \frac{1}{2}\overline{N} - \overline{m^2}) \sin^2 \theta, \tag{33a}$$

where $(\Delta N)^2$ is the variance about the mean total particle number $\overline{N}$ entering the interferometer, and $(\Delta m)^2$ is the variance about $\overline{m}$, which is one half the difference in mean numbers of particles $\overline{N_1}$ and $\overline{N_2}$ entering the two ports. Equation (33a) can also be written in the form

$$C(D_1, D_2) = C_{12} \cos^2 \theta + \frac{1}{4}((\Delta N)^2 - \overline{N} - 2\overline{N_1 N_2}) \sin^2 \theta, \tag{33b}$$

where

$$C_{12} = \overline{N_1 N_2} - \overline{N_1}\ \overline{N_2} \tag{33c}$$

is the cross-correlation in beam fluctuations at the input ports.

As a check of consistency, one can verify that if all the particles enter the interferometer through one of the ports—let us say port 1 (in which case $N = 2m$)—then upon substitution of $(\Delta N)^2 = 4(\Delta m)^2 = (\Delta N_1)^2$, the cross-correlation reduces to relation (25e) as expected.

Suppose, however, the input beam is in a pure SU(2) correlated state of precisely known particle number N and particle difference number m. Then the variances in N and m vanish, and the values $N_1 = (N/2) + m$ and $N_2 = (N/2) - m$ are also sharp. The cross-correlation in output counts reduces in that case to

$$C(D_1, D_2) = -\tfrac{1}{2}[N_1 N_2 + \tfrac{1}{2}(N_1 + N_2)] \sin^2 \theta, \tag{34a}$$

which is again intrinsically negative except at selected phase angles for which it vanishes.

In the more general case of an input beam comprising an appropriate mixture of SU(2) correlated states there is a range of phase angles, given by

$$\frac{4 C_{12}}{2\overline{N_1 N_2} + \overline{N} - (\Delta N)^2} > \tan^2 \theta, \tag{34b}$$

for which the cross-correlation can be positive. For example, at $\theta = 0$, the condition $(\Delta N)^2 > 4(\Delta m)^2$ results in a positive cross-correlation. This condition is most directly met by an ensemble of distributed total particle number, $(\Delta N)^2 > 0$, but sharp particle difference, $(\Delta m)^2 = 0$, at the input ports. An input composed of the states $|N/2, m\rangle$ and $|(N/2) + 2, m\rangle$ with $|m| \leq N/2$ leads to $C(D_1, D_2) = C_{12} = \tfrac{1}{4}$.

Consider next an ensemble described by a mixture of states of the form

$$|S(N)\rangle = \frac{1}{\sqrt{2}}(|N/2, \tfrac{1}{2}\rangle + |N/2, -\tfrac{1}{2}\rangle), \tag{35a}$$

where N is an odd integer and $(\Delta N)^2 > 1$. The analogy with angular momentum suggests that expression (35a) is a type of singlet state, hence the designation $|S(N)\rangle$. Since the cross-correlation

$$C(D_1, D_2) = \tfrac{1}{4}[(\Delta N)^2 - \cos^2 \theta] + \tfrac{1}{8}[(\Delta N)^2 - \tfrac{1}{2}(\overline{N} + 3)(\overline{N} - 1)]\sin^2 \theta, \tag{35b}$$

is positive for $(\Delta N)^2 > 1$ (which is characteristic of the proposed ensemble) and $\theta = 0$, the system provides another example of fermionic bunching.

It is unfortunate that there is at present no known way to produce these SU(2) correlated states in the laboratory, for not only do they manifest a rich and complex statistical behavior, but, from the perspective of interferometry, they could well be of practical use. A desirable objective is to enhance the sensitivity of an interferometer to the relative phase angle θ. In general, one would expect that the variance in θ, which can be calculated from expectation values of $N(D)$ as follows:

$$(\Delta\theta)^2 = \frac{(\Delta N(D))^2}{|\partial N(D)/\partial\theta|^2}, \tag{36a}$$

to diminish as some function of the number of particles—i.e., the larger the number of particles that pass through the interferometer, the more sharply is the relative phase angle θ determined. Ordinarily, this inverse dependence goes as the square root of the number of particles. For example, in the case of a beam described by a correlated state $|j, m\rangle$, (36a) reduces to

$$(\Delta\theta)^2 = \frac{1}{2}\left[\frac{N/2((N/2)+1)}{m^2} - 1\right] + \frac{(\Delta m)^2}{m^2}\cot^2\theta. \tag{36b}$$

Consider the special, but widely applicable, example of a single-port input of N particles described by the state $|N/2, N/2\rangle$. From relation (36b) it is seen that the *minimum* root-mean-square value of θ (for an uncorrelated input) is

$$(\Delta\theta)_{\text{min}}^{\text{unc.}} = \sqrt{\frac{N}{N^2}} \rightarrow \frac{1}{\sqrt{N}}, \tag{36c}$$

where the limiting expression is rigorously valid for sharp N. For the two-port "singlet" states of relation (35a), however, the variance in θ is given by

$$(\Delta\theta)^2 = \frac{4}{(N+1)^2} + \frac{(N+3)(N-1)}{(N+1)^2}\tan^2\theta, \tag{36d}$$

from which it follows that the expression comparable to (36c)

$$(\Delta\theta)_{\text{min}}^{S(N)} = \frac{2}{N+1} \tag{36e}$$

is inversely proportional to the number of particles (rather than the square root). Thus, the use of two-port correlated states can enhance the sensitivity of a fermion interferometer.

3.5. Brighter than a Million Suns: Electron Beams from Atom-Size Sources

The statistical properties of optical fields has been an active area of investigation for over thirty years, since stimulated by the pioneering experiments of Hanbury Brown and Twiss. As pointed out earlier, these experiments demonstrated correlations in intensity fluctuations and in photon arrival times of partially coherent light beams. In the parlance of optical coherence theory the HBT experiments manifest the *second-order* coherence characteristics of light deriving from wave noise (in the imagery of classical optics), or photon clumping (in the imagery of quantum optics).

This property of light is to be compared with *first-order* coherence which, both classically and quantum mechanically, refers to fringe contrast in an interference pattern.

The wave-particle duality intrinsic to quantum physics suggests that what can be done with light should also be feasible with massive particles, and, indeed, intensity interferometry has been used in nuclear physics to investigate, for example, the geometry of the emission region in high energy nuclear reactions [27]. The relativistic heavy ion beams employed in these experiments are basically incoherent, and the hadronic interactions to which they give rise are not sufficiently understood to permit theoretical calculation of the two-particle correlation functions. Such a source would not be suitable for quantum interference experiments of the kind described in the foregoing sections. What is needed is a bright, coherent, charged fermion source whose physical interactions are well understood. Practically speaking, this means an electron source. Weak interaction phenomena aside, the behavior of electrons (and positrons) is completely specified by quantum electro-dynamics. What are the prospects, then, of observing the quantum correlations of electrons with currently available electron sources?

It is well beyond the scope of this discussion to consider the details of all experimental difficulties that must be surmounted, but one conceptually important issue merits brief discussion. First and foremost, it is necessary to have a source for which there is a reasonable probability of obtaining correlated particles. Loosely speaking, the wave functions of two particles should in some sense overlap at the time of production (although it must be stressed again that classical images can be misleading, especially when employed to explain quantum effects of multiparticle states). All familiar electron sources—including the most coherent, such as the field-emission electron source—produce for the most part single-particle states. This is not simply a matter of beam *intensity*—i.e., a large electron accelerator in place of an electron microscope would not necessarily solve the problem—but rather a question of beam *brightness* which is related to the concept of degeneracy introduced earlier in (26b). It is worth examining more quantitatively the connections between the brightness, coherence, and degeneracy of a particle source.

Although the concept of brightness can be rigorously delineated [28], we adopt here the widely employed experimental definition of mean brightness B as the current density j (current per unit area normal to the beam) emitted into solid angle Ω. Thus, the number of particles received in a time interval Δt within a solid angle $\Delta \Omega$ through a detecting surface ΔA can be written as

$$\Delta n = (B/e) \, \Delta A \, \Delta t \, \Delta \Omega, \qquad (37a)$$

where e is the particle charge. The quantities preceded by Δ are to be regarded as sufficiently small (strictly speaking, infinitesimally small) so that particle direction, location, and arrival time are reasonably well defined. The beam degeneracy δ is defined in the alternative, but equivalent, expression based on the flux of unpolarized spin-$\frac{1}{2}$ particles with momentum $\mathbf{p}$ and speed v in phase space

$$\Delta n = (\text{mean number of particles per cell of phase space})$$
$$\times (\text{number of occupied cells})$$
$$= \delta\left(\frac{2\,\Delta p_x\,\Delta p_y\,\Delta p_z\,\Delta x\,\Delta y\,z}{h^3}\right) = \delta\left[\frac{2(p^2\,\Delta p\,\Delta\Omega)(v\,\Delta t\,\Delta A)}{h^3}\right]. \quad (37\text{b})$$

It is the degeneracy parameter which governs the performance of electron interferometers and fundamentally determines the magnitude of quantum interference effects involving electron correlations [29]. The factor 2 in (37b) takes account of the two spin degrees of freedom. Comparison of (37a) and (37b) leads to the expression

$$B = \delta B_{\max}, \quad (38\text{a})$$

where the maximum brightness

$$B_{\max} = 2evp^2\,\Delta p/h^3 = 2ep^2\,\Delta E/h^3 = 2e\,\Delta E/h\lambda^2, \quad (38\text{b})$$

occurs for degeneracy parameter $\delta = 1$. Conversion of momentum dispersion Δp into energy dispersion ΔE by use of the relativistically exact relation $E^2 = p^2c^2 + m^2c^4$ leads to the second equality in (38b). The third equality follows from substitution of the de Broglie relation $p = h/\lambda$ specifying the electron wavelength. For nonrelativistic electrons ($\Delta E = p\,\Delta p/m$) the maximum brightness can be conveniently expressed in terms of kinetic energy and energy dispersion

$$B_{\max}^{(\text{nonrel.})} = 4meE\,\Delta E/h^3. \quad (38\text{c})$$

Since the components of a split wave packet interfere to the extent that they occupy the same region of phase space, one can express the volume of a cell in phase space in terms of the (longitudinal) coherence length $l_c \sim vt_c$ and coherence area $A_c \sim l_t^2$ of the beam as follows:

$$(p^2\,\Delta p\,\Delta\Omega)(l_c A_c) = h^3, \quad (39)$$

where the coherence time t_c and coherence lengths l_c, l_t are defined in (1.2a, b, c). Use of relations (38a, b) and the definition of B then leads to an equivalent expression for degeneracy

$$\delta \sim (j/e)A_c t_c, \quad (40)$$

interpretable as the mean number of particles per coherence time traversing a coherence area normal to the beam. Relation (40) is essentially a re-expression of (26b), since the number of particles counted in a specified time interval (N_1/T) is equal to the particle flux through a coherence area ($j/e)A_c$ at the observation plane.

Experimentally, the brightness, and therefore the degeneracy, can be obtained directly from measurements of current density and solid angle or

Table I. Characteristics of a high-voltage field-emission electron beam.†

Energy, E	10^5 eV
Energy dispersion, ΔE	0.3 eV
Wavelength, λ	4×10^{-10} cm
Coherence time, t_c	1.4×10^{-14} sec
Current density, j	3.3×10^{13} e cm^{-2} s^{-1}
Angular divergence, α	1.3×10^{-7} rad
Transverse coherence length, l_t	1.5×10^{-3} cm
Coherence area, A_c	2.2×10^{-6} cm^2
Brightness, B	10^6 A cm^{-2} s^{-1} sr^{-1}
	6.3×10^{26} e cm^{-2} s^{-1} sr^{-1}
Degeneracy, δ	10^{-6}

† Adapted from Silverman [23].

indirectly from coherence parameters inferred from an interference pattern [30]. In the case of a coherent beam such as required for interferometry, the latter procedure is to be preferred. Of all the known types of particle sources, field-emission electron sources are the brightest currently available, considerably brighter than the Sun or linear accelerators or synchrotron/wiggler/undulator systems. For example, from knowledge of the solar constant (135 mW/cm^2 measured at the Earth over a spectral range of 330–1100 nm [31], the solar radius ($\sim 7 \times 10^8$ m), and the Earth–Sun distance ($\sim 1.5 \times 10^{11}$ m) one can estimate a solar brightness of approximately 10^{10} erg/s-cm^2-sr which translates into roughly 10^{21} photons/s-cm^2-sr, assuming a mean photon wavelength of 600 nm (yellow radiation). By constrast, as seen from Table I, the brightness of a standard 150 kV field-emission electron microscope is on the order of 10^8 amps/cm^2-sr or about 6×10^{26} electrons/s-cm^2-sr.

Nevertheless, a higher brightness does not, perforce, mean a greater degeneracy in comparisons between different types of sources. Unlike photon sources for which the degeneracy parameter can be many orders of magnitude—about 10^4 for an ordinary 1 mW HeNe laser, and in excess of 10^{16} for a ruby laser [32]—the maximum fermion degeneracy is unity. In practice, thermal neutron and thermal electron sources have a degeneracy parameter many orders of magnitude smaller than one, e.g., 10^{-12}–10^{-10}, and even in the case of a 150 kV field-emission source the degeneracy is only about 10^{-6}. It is the low fermion degeneracy that prompted Denis Gabor, inventor of holography, to note some thirty years ago "the almost complete identity of light optics and electron optics, in spite of the extreme difference between Einstein–Bose and Fermi–Dirac statistics" [33]. Still, the situation is improving.

A recent development that may augur well for the interferometry of correlated electrons is the production of ultrasharp field-emission tips

Figure 3.8. Configuration of a field-emission electron tip of atomic size. The spheres represent individual tungsten atoms from which electrons are preferentially emitted. (Adapted from Fink [34].)

emitting electrons from one or at most a few surface atoms [34] as schematically shown in Figure 3.8. Such "nanotips" are believed to produce beams brighter—and therefore more degenerate—by perhaps two or more orders of magnitude than previous field-emission tips [35].

All things being equal, the greater the degeneracy of the source, the more suitable is the source for use in experiments probing higher-order (than one) coherence effects. The degeneracy of the arc light source employed by HBT in experiments demonstrating photon correlations was about 10^{-3}. Although low, sufficient statistics were accumulated in approximately 10 h of counting to demonstrate the sought-for bosonic correlations. With nanometer-sized field-emission tips of brightness approximately 10^{10} A/cm²-sr one would have electron sources of degeneracy comparable to that of the light source employed by HBT. The observation of electron correlations may yet be difficult, but, Gabor's remark nothwithstanding, not outside the realm of possibility.

Given an appropriate source, at least three types of experimental approaches could in principle be employed to manifest electron correlations. One could count electrons arriving at a single detector to determine;

(a) the variance in the number of counts about the mean; or
(b) the conditional probability of receiving a second electron at a predetermined time interval after having detected a first.

Alternatively, one could correlate electron counts at two detectors so as to determine, for example,

(c) the number of coincidences as a function of optical path length difference.

Although there are practical differences in the implementation of the different

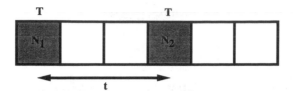

Figure 3.9. Sequence of time windows for an electron autocorrelation experiment. Electrons are counted for a period of T seconds in two windows separated by an interval t. Correlation of the fluctuations in counts about the mean value for each of the two windows reveals electron antibunching.

approaches, the expected ratio of signal (i.e., fermionic deviation from random particle statistics) to noise (the random background counts), does not differ substantially among them. We will consider as an illustration a simple, if somewhat idealized, experiment involving autocorrelation of electrons at a single detector.

Suppose that a beam of electrons is incident upon a single detector at a rate of $\mathscr{R}$ particles per second. One counts the number of particles received in a sequence of time slots or "windows" of duration T, and correlates the counts of windows separated by an adjustable time interval $t \geq T$ as shown schematically in Figure 3.9. The mean number of counts in two such windows would then be

$$\overline{N_1} = \overline{N_2} = \mathscr{R}T \qquad (41a)$$

and the correlation in fluctuations about the mean can be shown to be

$$\overline{\Delta N_1\, \Delta N_2} \sim -\overline{N_1}\ \overline{N_2}\frac{t_c}{t}, \qquad (41b)$$

where the coherence time of the source t_c is ordinarily much smaller than t. (The form of (41b) is similar to that of (5c) for photons, but with opposite sign.) Since the fermionic contribution to the fluctuations is very weak, it is the fluctuation in random arrival times (governed by Poisson statistics) that dominates the noise, in which case the variance in count rate in each time window is essentially equal to the count rate itself; consequently

$$\overline{(\Delta N_1)^2(\Delta N_2)^2} \approx \overline{(\Delta N_1)^2}\,\overline{(\Delta N_2)^2} \approx \overline{N_1}\ \overline{N_2}. \qquad (41c)$$

Hence, the signal-to-noise ratio $(S/N)_1$ for detecting fermionic correlations against the stochastic background counts for a single pair of windows and counting interval T is

$$(S/N)_1 = \frac{|\overline{\Delta N_1\, \Delta N_2}|}{\sqrt{\overline{(\Delta N_1)^2(\Delta N_2)^2}}} \sim (\overline{N_1}\ \overline{N_2})^{1/2}\frac{t_c}{t} = \mathscr{R}T\frac{t_c}{t}, \qquad (42a)$$

where the subscript 1 implies a single experimental run. By repeating the experiment n times to have a total counting time $T_{tot} = nT$ one enhances the signal-to-noise ratio by a factor $n^{1/2}$ to obtain

$$(S/N)_n = n^{1/2}(S/N)_1 \sim \delta \frac{\sqrt{T T_{\text{tot}}}}{t}. \tag{42b}$$

Note that the substituted relation $\delta = \mathcal{R} t_c$ is equivalent to (26b).

Expression (42b) can be rearranged to yield the total counting time required to achieve a desired (S/N).

$$T_{\text{tot}} = \frac{(S/N)^2 t^2}{\delta^2 T}. \tag{42c}$$

Thus, for a beam degeneracy $\delta \sim 4 \times 10^{-6}$, a beam coherence time $t_c \sim 10^{-14}$ s, a counting window $T = 5$ ns, and a delay interval $t = 5$ ns, one could ideally achieve a threshold signal-to-noise level $S/N = 1$ in under a half-hour ($T_{\text{tot}} \sim 26$ min). For S/N five times greater than threshold, a total time of approximately 10 hours—about the same as for the photon-counting experiment of HBT—is required to accumulate a sufficient number of counts. The total counting time could be reduced with shorter windows although this would necessitate faster responding particle detectors. In an actual experiment the T_{tot} may be lengthened considerably by constraints posed by additional data processing instrumentation. Conversely, the higher brightness of a nanotip source could conceivably increase the degeneracy by two orders of magnitude with a consequent reduction in counting time by four orders of magnitude.

In concluding this discussion of correlated particles, it is of conceptual interest to reconsider the fundamental criterion, first enunciated by Dirac [36], for the occurrence of particle interference. Expressed originally in regard to light, but equally relevant to the behavior of all matter, Dirac declared: "Each photon (in a split beam) interferes *only* with itself. Interference between different photons *never* occurs." (Italics added.) What distinguishes the quantum interference processes described in this chapter from those of which Dirac had in mind is precisely that they (the former) are intrinsically *multiparticle* phenomena. Do these new kinds of interferometry—in which quantum statistics and long-range correlations play seminal roles—refute Dirac's dictum? I do not think this is the case. Rather, quantum mechanics redefines more subtly what must be understood as "different" particles. We have seen in the examples considered that particles, whether correlated by the spin-statistics connection or by the initial conditions of their production, are indistinguishable parts of a single system however far apart they may subsequently be at the moment of observation. Under circumstances where interference occurs, one cannot tell from which part of the source a given particle comes, or to which detector it goes. It is then operationally meaningless to speak of interference between "different" particles. What one might say instead, is that "a system interferes only with itself."

References

[1] A. Einstein, B. Podolsky, and N. Rosen, Can Quantum Mechanical Description of Physical Reality Be Considered Complete?, *Phys. Rev.*, **47**, 777–780 (1935).

[2] N. Bohr, Can Quantum Mechanical Description of Physical Reality Be Considered Complete?, *Phys. Rev.*, **48**, 696–702 (1935).

[3] M. Jammer, *The Philosophy of Quantum Mechanics* (Wiley, New York, 1974), p. 187.

[4] See, for example, the comprehensive technical discussion by J.F. Clauser and A. Shimony, Bell's Thorem: Experimental Tests and Implications, *Rep. Progr. Phys.* **41**, 1881–1927 (1978); and the popular articles by B. d'Espagnat, The Quantum Theory and Reality, *Scientific American*, **247**, 158–181 (November, 1979); T.A. Heppenheimer, Experimental Quantum Mechanics, *Mosaic* **17**, 19–27 (1986); and A. Shimony, The Reality of the Quantum World, *Scientific American*, **256**, 46–53 (January, 1988).

[5] R. Hanbury Brown and R.Q. Twiss, Correlation between Photons in Two Coherent Beams of Light, *Nature*, **177**, 27–29 (1956).

[6] R. Hanbury Brown, *The Intensity Interferometer* (Taylor and Francis, New York, 1974), p. 7.

[7] E.M. Purcell, The Question of Correlation between Photons in Coherent Light Rays, *Nature*, **178**, 1449–1450 (1956).

[8] A. Einstein, Zum gegenwärtigen Stand des Strahlungsproblems (On the Current State of the Radiation Problem), *Phys. Zeit.*, **10**, 185–193 (1909).

[9] For a comprehensive description of the nature of chaotic light, which includes black-body radiation as a special case, see R. Loudon, *The Quantum Theory of Light*, 2nd edn. (Oxford, New York, 1983), pp. 157–160. A significant feature is that the density or statistical operator of a chaotic radiation field is diagonal in a basis of photon number states. For example, in the case of a single optical mode, it would have the form

$$\hat{\rho} = \sum_{\{n\}} \rho_{n,\,n} |n\rangle\langle n|.$$

[10] See, for example, the proceedings of the XXth Solvay Conference on Physics: Quantum Optics, edited by P. Mandel, *Phys. Rep.* **219** (North-Holland, Amsterdam, 1992).

[11] M.P. Silverman, Applications of Photon Correlation Techniques to Fermions, *OSA Proceedings on Photon Correlation Techniques and Applications*, Vol. 1, edited by J. B. Abbiss and A. E. Smart (OSA, Washington, DC, 1988), pp. 26–34.

[12] M.P. Silverman, Second-Order Temporal and Spatial Coherence of Thermal Electrons, *Nuovo Cimento B*, **99**, 227 (1987).

[13] M.P. Silverman, Two-Solenoid Aharonov–Bohm Experiment with Correlated Particles, *Phys. Lett. A*, **148**, 154 (1990).

[14] M.A. Horne, A. Shimony, and A. Zeilinger, Two-Particle Interferometry, *Phys. Rev. Lett.*, **62**, 2209 (1989).

[15] A. Zeilinger, General Properties of Lossless Beam Splitters in Interferometry, *Amer. J. Phys.*, **49**, 882 (1981).

[16] M.P. Silverman, More Than One Mystery: Quantum Interference with Correlated Charged Particles and Magnetic Fields, *Amer. J. Phys.*, **61**, 514 (1993).

[17] M.P. Silverman, New Quantum Effect of Confined Magnetic Flux on Electrons, *Phys. Lett. A*, **118**, 155 (1986).

[18] M.P. Silverman, Quantum Interference Effects on Fermion Clustering in a Fermion Interferometer, *Physica B*, **151**, 291 (1988).

[19] The coherent states $|\alpha\rangle$ of a single-mode oscillator (from which model the optical states are derived) can be expressed in a basis of energy (or excitation number) states as follows:

$$|\alpha\rangle = \exp(-|\alpha|^2/2) \sum_{n=0}^{\infty} \frac{\alpha^n}{\sqrt{n!}} |n\rangle.$$

For the properties of coherent states and a detailed exposition of photon statistics, see R.J. Glauber, Optical Coherence and Photon Statistics, in *Quantum Optics and Electronics*, edited by C. DeWitt et al. (Gordon & Breach, New York, 1965), pp. 65–185.

[20] See [10], pp. 226–229.

[21] M.P. Silverman, Fermion Ensembles That Show Statistical Bunching, *Phys. Lett. A*, **124**, 27–31 (1987).

[22] An introduction to the application of second quantization to light is given by D.F. Walls, A Simple Field Theoretic Description of Photon Interference, *Amer. J. Phys.*, **45**, 952–956 (1977).

[23] M.P. Silverman, On the Feasibility of Observing Electron Antibunching in a Field-Emission Beam, *Phys. Lett. A*, **120**, 442–446 (1987).

[24] M.P. Silverman, Gravitationally Induced Quantum Interference Effects on Fermion Antibunching, *Phys. Lett. A*, **122**, 226–230 (1987).

[25] R. Colella, A.W. Overhauser, and S.A. Werner, Observation of Gravitationally Induced Quantum Interference, *Phys. Rev. Lett.*, **34**, 1472–1474.

[26] B. Yurke, Input States for Enhancement of Fermion Interferometer Sensitivity, *Phys. Rev. Lett.*, **56**, (1986) 1515–1517.

[27] D.H. Boal, C-K. Gelbke, and B.K. Jennings, Intensity Interferometry in Subatomic Physics, *Rev. Mod. Phys.*, **62**, 553–602 (1990).

[28] P. Hawkes and E. Kasper, *Electron Optics*, Vol. 2 (Academic Press, New York, 1989), p. 271.

[29] M.P. Silverman, Distinctive Quantum Features of Electron Intensity Correlation Interferometry, *Nuovo Cimento B*, **97**, 200 (1987).

[30] J.C.H. Spence, W. Qian, and M.P. Silverman, Electron Source Brightness and Degeneracy from Fresnel Fringes in Field Emission Point Projection Microscopy, *J. Vac. Sci. Technol. A*, **12**, 542–547 (1994).

[31] J.M. Pasachoff, *Contemporary Astronomy* (W.B. Saunders, Philadelphia, 1977), pp. 167–168.

[32] The degeneracy of a quasi-monochromatic laser source of power P, frequency v, and bandwidth Δv or pulse width τ is effectively the number of photons $\delta = Pt_c/hv$ emitted in a coherence time $t_c \sim 1/\Delta v$ or τ. Thus, a continuous-wave HeNe beam of wavelength 633 nm and spectral width 0.2 nm can be shown to have a degeneracy of approximately 2.14×10^4. A ruby laser producing a train of 5 mW pulses each of 1 μs duration at 694 nm emits about 1.8×10^{16} photons per pulse. See, for example, B. Lengyel, *Lasers* (Wiley, New York, 1971), p. 138.

[33] D. Gabor, Light and Information, in *Progress in Optics*, Vol. 1, edited by E. Wolf (North-Holland, Amsterdam, 1961), pp. 109–153 (quotation from p. 148).

[34] H.W. Fink, Point Source for Ions and Electrons, *Phys. Scripta*, **38**, 260–263

(1988); P.A. Serena, L. Escapa, J.J. Saenz, N. Garcia, and H. Rohrer, Coherent Electron Emission from Point Sources, *J. Microscopy*, **152**, 43–51 (1988).

[35] W. Qian, M.R. Scheinfein, and J.C.H. Spence, Brightness Measurements of Nanometer-sized Field-Emission-Electron Sources, *J. Appl. Phys.* **73**, 7041 (1993).

[36] P.A.M. Dirac, The Principles of Quantum Mechanics, 4th edn. (Oxford, London, 1958), p. 9.

CHAPTER 4

Quantum Boosts and Quantum Beats

Keeping time, time, time
In a sort of runic rhyme,
To the tintinnabulation that so musically wells
From the bells, bells, bells, bells
bells, bells, bells—
From the jingling and the tinkling of the bells.

The Bells
E.A. Poe

4.1. Interfering Pathways in Time

We have seen in preceding chapters that the potential for quantum interference exists whenever a particle can propagate from its source to the detector by alternative spatial pathways under experimental conditions such that the exact pathway taken cannot be known. The archetypal example is the Young's two-slit experiment in which the particle, when probed, passes through one slit or the other. Unprobed, the resulting particle distribution is explicable only in terms of probability amplitudes that seemingly propagate through both slits. There is a direct temporal analogue to the two-slit experiment in which the linearly superposed amplitudes represent—not alternative spatial pathways—but rather the evolution of alternative indistinguishable events in time.

Of the many ways in which the structure of atoms, molecules, and other bound-state quantum systems differs from that of macroscopic classical systems, one of the most striking is the discreteness or quantization of energy. The internal energy of a classical planetary system is an accident of its formation and likely to differ from one system to another even though the two systems may be composed of identical masses. By contrast, every ground-state hydrogen atom has the same energy irrespective of its formation; likewise for identical atoms in corresponding excited states.

Although the idea of electrons populating quantized energy eigenstates is a familiar one, it is nevertheless necessary to be careful lest uncritical

100

usage provide a misleading picture of the atom and its interactions. The Russian spectroscopist E.B. Aleksandrov expressed this point very well when he wrote [1]:

In connection with the remarkable progress made in the interpretation of atomic spectra, the concepts of energy levels and their populations have become so firmly entrenched in atomic physics and spectroscopy that they became gradually independent concepts, losing the meaning attributed to them by quantum mechanics. Yet the statement commonly made in spectroscopy, that the atom is at a given (excited) level, is incorrect in the overwhelming majority of cases.

In such circumstances the energy of an atom is actually indeterminate, the atomic state being represented by a linear superposition (with appropriate amplitudes) of all possible stationary states that can be reached by the particular type of excitation employed. One optical consequence of this coherent superposition is that the rate of spontaneous emission (i.e., the fluorescence intensity) from an ensemble of atoms excited in this manner can oscillate in time while diminishing exponentially, such as illustrated in Figure 4.1 for the case of a four-level atom with two closely spaced excited states. For these "quantum beats" to be observable, the emission events among the various atoms must be in some sense synchronized; otherwise the superposition of oscillating intensities widely out of phase would display no net modulation. Had the atoms decayed from well-defined energy states, the temporal variation in fluorescence would have been

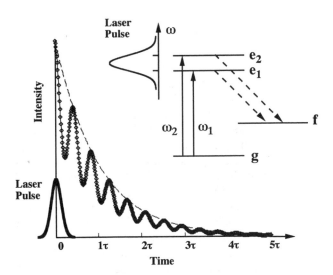

Figure 4.1. Coherent excitation of two close-lying excited states by a pulsed laser giving rise to oscillations (quantum beats) in the spontaneous emission as a function of time. A light pulse narrow in the time domain has a broad frequency spectrum which, shown in the upper insert, contains Fourier components capable of inducing transitions from the ground state to both excited states.

strictly an exponential decay. Thus, there is a significant conceptual and experimental distinction between a quantum ensemble designated a *mixture* of states, wherein each constituent is endowed with definite, although statistically distributed energy values, and another ensemble designated a *superposition* of states, with each constituent in a linear superposition of energy eigenstates encompassing the same energy values and populations as the first ensemble [2].

Production of an atom (or molecule) in a linear superposition of excited energy states ordinarily requires an impulsive excitation ("quantum boosts"), i.e., a process of sufficiently short duration that its Fourier spectrum contains frequency components corresponding to the energy intervals between ground and excited states. Thus, for the example shown in the insert of Figure 4.1, the spectral width of the excitation must satisfy the relation

$$\Delta\omega \geq \omega_0 \equiv (E_2 - E_1)/\hbar = \omega_2 - \omega_1, \tag{1}$$

where ω_0 is the Bohr angular frequency of the excited level. Although the theoretical feasibility of such light oscillations was discussed in papers shortly following the creation of quantum mechanics [3], the actual experimental implications were not conceptually appreciated or experimentally realized until some thirty years later [4, 5, 6] principally by those involved with optical pumping. (Optical pumping refers to the use of light to populate a set of energy levels with a distribution different from that of a normal Boltzmann distribution at the temperature of the experiment [7]). There are a variety of ways of achieving the impulsive excitation required to generate a superposition state and the ensuing modulated fluorescence, as, for example, by light pulses [8, 9], pulsed electron impact [10], and electron capture collisions with a thin carbon foil target [11].

Like the quantum intereference phenomena described in earlier sections, the phenomenon of quantum beats under discussion is intrinsic to each atom and not a cooperative interaction between atoms. In other words, the spontaneous emission from single atoms is *not* modulated, but registers at the detector as one quantum of light at a time; the pattern of beats (measured at one location in real time or, equivalently, at different spatial locations along an accelerated atomic beam) can nevertheless be built up by the decay of many such single atoms. This is again the old "mystery" of quantum interference translated to the time domain: How can independently excited, randomly decaying, noninteracting atoms produce a pattern of photon arrivals that oscillates in time? Note that the synchronization required for the beats to survive ensemble averaging does not imply that emitting atoms communicate with or influence one another. Rather, an apt analogy, if there be any, would be that of a large number of independent clocks all separately wound and set to the same time by the clockmaker.

There is no classical mechanism for the interference, but quantum mechanics does allow one to analyze the phenomenon mathematically. Consider, again, the system illustrated in Figure 4.1 in which it is assumed for simplicity that the two excited states have the same lifetime $\tau = 1/\Gamma$ where Γ is the decay rate. Let a_{gi} be the amplitude for a transition from the ground state g to excited state e_i ($i = 1, 2$), and b_{if} be the corresponding amplitude for radiative decay from e_i to the lower state f (which could be the ground state). Then, the net amplitude for excitation to state e_i and subsequent decay after a time interval t takes the form

$$A_{g-i-f}(t) = a_{gi} \exp(-iE_i t/\hbar) e^{-\Gamma t} b_{if} \equiv a_{g-i-f} \exp(-iE_i t/\hbar) e^{-\Gamma t/2}, \quad (2a)$$

where $a_{g-i-f} = a_{gi} b_{if}$. We have made the heuristic assumption—to be examined further in the next section—that the excitation occurs effectively instantaneously. If the energy dispersion of the excitation mechanism is sufficiently broad, and quantum selection rules do not forbid the pertinent transitions, there are two indistinguishable pathways in time by which the atom can pass from the initial level g to the final level f; the corresponding total amplitude for the process is

$$A_{g \to f}(t) = [a_{g-1-f} \exp(-iE_1 t/\hbar) + a_{g-2-f} \exp(-iE_2 t/\hbar)] e^{-\Gamma t/2}. \quad (2b)$$

Hence, the probability for the transition to occur at time t with emission of one quantum of light—although from which state the observer does not know—is

$$P_{g \to f}(t) = |A_{g \to f}(t)|^2 = [|a_{g-1-f}|^2 + |a_{g-2-f}|^2 + 2a_{g-1-f} a_{g-2-f} \cos(\omega_0 t)] e^{-\Gamma t}. \quad (2c)$$

For purposes of illustration, the amplitudes a_{g-i-f} were assumed to be real valued. If, by some means—e.g., by placing an optical narrow-band filter before the detector—the observer can select only photons of energy E_1 or E_2 and thereby ascertain the temporal pathway by which the system evolved, the quantum beats would disappear and the system would simply decay exponentially in time.

As a spectroscopic method, the observation of quantum beats has a number of significant advantages compared with alternative procedures that probe the atom with resonant oscillatory fields. For one thing, the presence of an external oscillating field, as will be shown later in Chapter 5, can affect the level separation and decay of the states being examined. With quantum beat spectroscopy the atoms "ring out" their level structure without being probed. Since the strong impulsive fields that produce the excited states can be made to vanish substantially by the time the atoms are likely to decay, one can investigate a sample of interest in a resonance cell (or "bottle)— rather than employ the more complicated technology of an accelerated beam—and still be able to separate the processes of excitation and spectroscopy. Second, even though the atoms in a gas or vapor may be moving randomly about the interior of a resonance cell, rather than moving with well-defined linear momentum along a beam, the optical *signal*—i.e., the low-frequency modulation, as opposed to the high-frequency optical carrier wave—is largely free of Doppler broadening. Since the beat frequency is proportional to the difference in energy of two levels of the *same* atom, the Doppler shifts of the photons <u>potentially emissible</u> from each level nearly cancel. The underscored words again emphasize the fact that under the given circumstances only one quantum or light, not two, actually emerges from the spontaneous decay of a single atom. The beat is intrinsic to each atom, but made manifest only by the decay of a large number of similarly prepared atoms.

4.2. Laser-Generated Quantum Beats

Of the many ways to excite an atom impulsively into a linear superposition of states, one of the most practically useful and conceptually interesting is the application of pulsed laser light [12, 13]]. The use of light in general—as opposed to impulsive excitation by some form of particle bombardment, for example—is most amenable to theoretical analysis, since the interaction of matter with light is completely accounted for by quantum electrodynamics, undoubtedly the most thoroughly understood of all physical interactions. (By contrast, coherent atomic excitation by particle bombardment is not as simply analyzable or well understood.) It was, in fact, by means of pulsed light from shuttered spectral (i.e., nearly monochromatic) lamps that the phenomenon of field-free quantum beats was first observed. These sources, however, were weak in the sense that the probability of atomic excitation was low; at any moment the likelihood was greatest to find the illuminated atom in its ground state.

The development of tunable pulsed lasers has made possible the high intensity, short pulse duration, and broad spectral width that are advantageous in the study of some intriguing, if not outright exotic, physical systems such as Rydberg atoms [14]—i.e., atoms so highly excited that they begin to resemble in many ways (but not all [15]) minute classical planetary systems. The wide range over which such lasers can be tuned allows one to excite a large number of UV, visible, and IR transitions. The high power makes it possible to saturate even weakly allowed transitions. And when, because of parity restrictions, a single laser cannot induce transitions between two states of interest, two or more lasers used sequentially can effect the desired result by a stepwise excitation. These three advantages, besides those already cited common to quantum beat spectroscopy irrespective of the method of excitation, permit an experimenter to select almost any Rydberg state of interest or to study with facility an entire series of states [16].

It is the high intensity of pulsed lasers, however, that enriches (or complicates—depending on one's point of view!) the interaction between the atoms and light. The weak-pumping approximation implicit in the simple analysis of Section 4.1 is, strictly speaking, a first-order perturbation calculation in which the atomic system interacts at most once with the exciting light. In other words, the atom is presumed to absorb a single photon during the passage of the pulse (and to emit a photon once after the pulse has passed). Thus, this mode of treatment is also known as the linear absorption approximation. While valid for classical light sources, the linear absorption approximation is no longer *a priori* justified when the light source is a high-power pulsed laser. For one thing, while illuminated by the light pulse, a particular atom can be driven back and forth a number of times between the ground and excited states by the processes of photon absorption followed by stimulated emission. In addition, if the spectral profile of the exciting light is not symmetric, or if it is not centered precisely on the transition to be effected, the light could displace the atomic energy levels from their vacuum values. These processes, absent under conditions of weak pumping, can modify the amplitude and phase of the ensuing quantum beats—sometimes in an extraordinary way as will be discussed shortly.

The theory of light-induced quantum interference that will be discussed in this section is valid under the relatively nonrestrictive limitation to *broad-band* pulse excitation. This condition, realized in most quantum-beat experiments, permits the theory to be formulated within the framework of the classical optical pumping cycle developed by Barrat and Cohen-Tannoudji [17] for weak light sources and subsequently generalized by others for continuous lasers [18].

Let us generalize somewhat the simple atomic structure assumed in Section 4.1 by considering an atom with three groupings (or manifolds) of states:

(i) ground g;
(ii) excited e; and
(iii) final f;

where each manifold can contain more than one state. In a quantum beat experiment envisioned here the atoms are excited from the g to e manifold by means of a light pulse of polarization ε and duration T and subsequently decay optically at rate Γ to the lower manifold f. (Different excited states could have different lifetimes, but this would unnecessarily complicate the conceptual ideas to be elucidated here.) After passage of the pulse, fluorescence of a particular polarization $\varepsilon_{\mathbf{d}}$ is observed. In order that the preparation of the excited states be well separated in time from the detection of the spontaneous emission, it is necessary that T be short in comparison to the lifetime $\tau = 1/\Gamma$.

The existence of a multiplicity of states (at least two) in the e-manifold is essential to the production of quantum beats. While substructure in the lower g and f manifolds is not essential, such structure can affect the phase and amplitude of the beats. Similarly, the energy resolution of the detector—and therefore the number of final states involved in the decay process—may also influence the beat signal. It is of interest to note that a semiclassical theory of radiative phenomena (termed the neoclassical theory) proposed some years ago [19] leads to quantum beats from single atoms with one excited state and a multiplicity of lower states. This is forbidden by quantum electrodynamics (QED) which— excluding processes involving the weak nuclear interactions (discussed in Chapter 6)—remains unchallenged within its domain of validity by any reliably reproducible experiment. QED does, however, permit quantum beats arising from lower state splittings in the case of *cooperative* interactions between two or more identical atoms [20].

In contrast to the heuristic linear-approximation analysis of Section 4.1, which determines the net *amplitude* for transition from the g to the f manifolds, the optical pumping equations determine the atomic *density matrix* $\rho(t)$, which enters directly into the calculation of the mean value of an observable (represented by operator $\mathcal{O}$) through the trace relation

$$\langle \mathcal{O} \rangle = \text{Tr}\{\rho\mathcal{O}\}. \tag{3}$$

The elements of the density matrix are ensemble-averaged bilinear combinations of transition amplitudes.

In determining the time evolution of a quantum system subject to an external interaction, it is often convenient to eliminate at the outset from the equations of motion the time-dependence arising from the internal interactions governed by the field-free Hamiltonian H_0, since this evolution is already known and generally involves the highest frequencies. To do this, one transforms the equations of motion into the interaction representation. The transformed density matrix

$$\bar{\rho}(t) = \exp(iH_0t/\hbar)\rho(t)\exp(-H_0t/\hbar) \tag{4}$$

is independent of t before and after the passage of the light pulse when the external interaction vanishes. Let ρ_- and ρ_+ characterize the atom at the temporal limits $t \rightarrow -\infty$ and $t \rightarrow +\infty$, respectively. The effect of the laser pulse is then entirely known if one can determine the time evolution of ρ_- into ρ_+.

The time evolution of the atomic density matrix can be determined from a set of optical pumping equations generalized to include pulsed laser excitation [12, 13]. Although the derivation of these equations will not be given here, the structure of the equations and the assumptions underlying the derivation will be discussed. First, one decomposes the atomic density matrix into a submatrix of excited states ρ_e and of ground states ρ_g by means of projection operators P_e and P_g in the following standard way:

$$\rho_e = P_e\rho P_e, \qquad \rho_g = P_g\rho P_g, \tag{5a}$$

where each projection operator has the form

$$P_\mu = \sum_i |\mu_i\rangle\langle\mu_i| \tag{5b}$$

with the summation extending over the states of the appropriate manifold ($\mu = e$ or g). This leads to the following system of equations coupling the elements of ρ_e and ρ_g:

$$d\rho_e/dt = -(i/\hbar)[H_0, \rho_e] - \tfrac{1}{2}\{\Gamma_e, \rho_e\} + (1/T_p)P_e\varepsilon \cdot \mathbf{D}P_g\rho_g P_g\varepsilon^* \cdot \mathbf{D}P_e$$
$$- (\tfrac{1}{2}T_p)\{P_e\varepsilon \cdot \mathbf{D}P_g\varepsilon^* \cdot \mathbf{D}P_e, \rho_e\}$$
$$- (i\Delta E(t))[P_e\varepsilon \cdot \mathbf{D}P_g\varepsilon^* \cdot \mathbf{D}P_e, \rho_e], \tag{6a}$$

$$d\rho_g/dt = -(i/\hbar)[H_0, \rho_g] - (\tfrac{1}{2}T_p)\{P_g\varepsilon^* \cdot \mathbf{D}P_e\rho_g P_g\varepsilon \cdot \mathbf{D}P_g, \rho_g\}$$
$$+ (1/T_p)P_g\varepsilon^* \cdot \mathbf{D}\rho_e P_e\varepsilon \cdot \mathbf{D}P_g,$$
$$+ (i\Delta E(t))[P_g\varepsilon^* \cdot \mathbf{D}P_e\varepsilon \cdot \mathbf{D}P_g, \rho_g], \tag{6b}$$

Although seemingly complicated, the various terms of the above equations (where $[x, y]$ represents the commutator and $\{x, y\}$ the anticommutator of two operators x and y) are amenable to simple interpretation. The first term of (6a) and (6b) involving the commutator with H_0 represents the free evolution in time of the two sets of states. The other terms depending only on the elements of the ground state matrix ρ_g represent the effects of optical absorption. This process in (6a) populates the excited states at a rate proportional to the reciprocal of the so-called pumping time $T_p(t)$ which is, itself, inversely related to the strength of the exciting pulse in the following way:

$$1/T_p(t) = (\pi/\hbar^2)\mathscr{F}(t, \omega_{eg})|(D_r)_{eg}|^2. \tag{7a}$$

Here $\mathscr{F}(t, \omega)$ is the spectral density of the electric field $\mathbf{E}$ of the laser pulse, i.e., the Fourier transform (at angular frequency ω) of the electric field auto-correlation function

$$\mathscr{F}(t, \omega) = \int \langle \mathbf{E}(t)\mathbf{E}^*(t - t')\rangle e^{i\omega t'}\, dt'. \tag{7b}$$

The transform in relation (7a) is evaluated at the Bohr frequency ω_{eg} characterizing the energy interval betweeen the ground and excited states. The factor

$$(D_r)_{eg} \equiv \langle e|D_r|g\rangle \tag{7c}$$

is the radial matrix element of the electric dipole operator $\boldsymbol{\mu}_E = D_r\mathbf{D}$ here written as the product of a scalar radial part D_r and vectorial angular component $\mathbf{D}$. Absorption also serves in (6b) to depopulate the ground states at a rate proportional to $1/2T_p(t)$ and to displace the different substates of g by an amount $\Delta E(t)$ where

$$\Delta E(t) = (1/\hbar)|(D_r)_{eg}|^2\mathscr{P}\int_{-\infty}^{\infty} \frac{\mathscr{F}(t, \omega)}{\omega - \omega_{eg}}\, d\omega. \tag{7d}$$

The symbol $\mathscr{P}$ in (7d) represents the Cauchy principal value of the integral.

Correspondingly, except for the second term of (6a) which is responsible for state decay by spontaneous emission, all other terms in the optical pumping equations depending on the elements of the excited state density matrix ρ_e represent the effects of stimulated emission. This process is symmetrical to that of absorption; it serves to populate the ground states in (6b) and to depopulate and displace the excited states in (6a). The process of spontaneous emission in (6a) is accounted for by an anticommutation operation with a phenomenological decay operator Γ_e which takes the form (in an energy representation) of a diagonal matrix whose elements are the decay rates (inverse lifetimes) of the excited states. This procedure is justifiable by rigorous application of quantum electrodynamics [21] and is applicable even if the atomic system is subjected to external fields [22]. To a good approximation the theoretical effect of decay is to multiply the atomic density matrix ρ_+ in the absence of spontaneous emission by the factor $\exp(-\Gamma_e t)$.

In the derivation of (6a) and (6b), it has been assumed that the correlation time t_c of the light pulse—i.e., the inverse of the spectral width Δ—is much shorter than the pulse duration T, i.e.,

$$t_c = 1/\Delta \ll T. \tag{8a}$$

The correlation time is a measure of the time interval over which the phase of the electric field is well defined. The further assumption that

$$t_c \ll T_p \tag{8b}$$

signifies that the phase of the incident pulse undergoes many random fluctuations over the period required to pump an atom out of its ground state. Equation (8b) places an upper limit on the pulse intensity—since the stronger the pulse, the shorter the pumping time—beyond which the optical pumping equations may no longer be valid. The above two conditions are not too restrictive, however, and are generally satisfied in standard quantum beat experiments. For example, a dye laser of a few hundred watts peak power yields temporal parameters on the order of

Pulse width	Pumping time	Coherence time
$T \sim 10^{-9}$ s $>$	$T_p(t) \sim 10^{-10}$ s $>$	$t_c \sim 10^{-11}$ s

for a well-allowed pumping transition.

The assumption has also been made in deriving (6a), (6b) that the Bohr frequencies in the manifolds e and g are small in comparison to Δ. This simplifies the analysis by leading to a unique pumping time and level shift for all the substates of a given manifold, but is not essential to the validity of the basic theoretical approach. The added complication of T_p and ΔE varying with the states of a manifold can be incorporated into the theory whenever necessary.

It is to be noted that no density matrix elements of the form ρ_{eg} or ρ_{ge}, which characterize optical coherence terms, appear in the optical pumping equations. These contributions—interpretable according to classical imagery as a macroscopic electric dipole moment precessing at optical frequencies—vanish when averaged over many correlation times of the field. In the broad-band approximation, as the above theoretical approach is known, all relevant time parameters are *long* in comparison to the correlation time.

Solving the equations of motion and implementing the appropriate initial conditions lead to the matrix ρ_+ and, from (4), to the expression

$$\rho_e(t) = P_e \rho(t) P_e = P_e \exp(-iH_0 t/\hbar)\rho_+ \exp(iH_0 t/\hbar)P_e\, e^{-\Gamma_e t} \qquad (9)$$

for the total atomic density matrix projected onto the excited states. The optical signal, observed in the spontaneous emission after the pulse excitation is concluded, is obtained from relation (3)

$$I(\varepsilon_\mathbf{d}) = K\,\mathrm{Tr}\{\rho_e(t)\mathcal{O}^\dagger_{\mathrm{det}}(\varepsilon_\mathbf{d})\}\, e^{-\Gamma_e t}, \qquad (10a)$$

where the detection operator $\mathcal{O}_{\mathrm{det}}(\varepsilon_\mathbf{d})$ describes the optical transition (with polarization $\varepsilon_\mathbf{d}$) between the e and f manifolds

$$\mathcal{O}_{\mathrm{det}}(\varepsilon_\mathbf{d}) = P_e \varepsilon_\mathbf{d} \cdot \mathbf{D} P_f \varepsilon_\mathbf{d}^* \cdot \mathbf{D} P_e. \qquad (10b)$$

(Note that only the angular component of the electric dipole operator appears in (10b); the radial part has already been incorporated in the definition of the pumping time T_p.) The constant K depends on geometrical factors such as the solid angle of acceptance of the detector and distance of the detector from the decaying atoms. As a scaling factor, it does not affect the form of the quantum beat signal and will henceforth be disregarded.

From the mathematical form of (10a) one clearly sees that superposed on the exponential decay are modulations of the fluorescence at the various Bohr frequencies of the excited manifold. The amplitude of each Fourier component of the beat signal depends on the elements of $\mathcal{O}_{\mathrm{det}}(\varepsilon_\mathbf{d})$ and ρ_+. To illustrate the use and physical content of the optical pumping equations—with an eye toward the specific effects of nonlinearity in the interaction between the atoms and the light pulse—we will examine in the next section the quantum beat signals produced by a three-level system subjected to different excitation conditions.

4.3. Nonlinear Effects in a Three-Level Atom

Consider an atom with nondegenerate ground state g and two excited states e_1 and e_2 both of which decay at the rate Γ. This system is almost the same as that treated in Section 4.1, except that states g and f are here taken to be identical. The processes of excitation and decay, summarized in Figure 4.2, are characterized by the respective matrix elements

$$a_j \equiv \langle e_j|\varepsilon\cdot\mathbf{D}|g\rangle \qquad (j = 1, 2), \qquad (11a)$$

$$b_j \equiv \langle e_j|\varepsilon_\mathbf{d}\cdot\mathbf{D}|g\rangle \qquad (j = 1, 2), \qquad (11b)$$

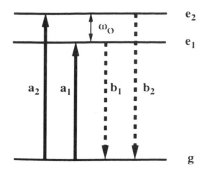

Figure 4.2. Nondegenerate three-level system showing amplitudes for excitation and decay processes.

which, for the sake of simplicity, are assumed to be real. The matrix representation of the detection operator then takes the form

$$\mathcal{O}_{\text{det}} = \begin{pmatrix} b_1^2 & b_1 b_2 \\ b_1 b_2 & b_2^2 \end{pmatrix}. \tag{12}$$

There are four independent elements to the density matrix of the atomic system which are designated as follows:

$$
\begin{aligned}
x &\equiv \langle g|\rho|g\rangle, \\
y_j &\equiv \langle e_j|\rho|e_j\rangle \quad (j = 1, 2), \\
z &\equiv \langle e_2|\rho|e_1\rangle = \langle e_1|\rho|e_2\rangle^*,
\end{aligned} \tag{13a}
$$

where the initial conditions before the laser excitation

$$x_- = 1, \qquad y_{1-} = y_{2-} = z_- = 0 \tag{13b}$$

characterize a ground-state atom. The corresponding density matrix elements characterizing the state of the atom after passage of the pulse will be subscripted with "+." The quantum beat signal, (10a) then takes the form of (2c)

$$I = [b_1^2 y_{1+} + b_2^2 y_{2+} + b_1 b_2(z_+ + z_+^*) \cos(\omega_0 t)]e^{-\Gamma t}, \tag{14}$$

where, as before, ω_0 is the angular frequency corresponding to the excited state energy interval.

To calculate the signal intensity I explicitly one must solve the optical pumping equations

$$dx/dt = \frac{1}{T_p(t)} [a_1^2 y_1 + a_2^2 y_2 + a_1 a_2(z + z^*) - (a_1^2 + a_2^2)x], \tag{15a}$$

$$dy_j/dt = \frac{1}{T_p(t)} [a_j^2(x - y_j) - \frac{a_1 a_2}{2}(z + z^*)] \pm i a_1 a_2 \, \Delta E(t)(z^* - z), \tag{15b}$$

with $\pm$ referring to $j = 1, 2$, respectively, and

$$dz/dt = \frac{1}{T_p(t)}\left[a_1 a_2\left(x - \frac{y_1 + y_2}{2}\right) - \frac{a_1^2 + a_2^2}{2}z\right] - i\omega_0 z$$
$$- i\Delta E(t)[a_1 a_2(y_1 - y_2) + (a_2^2 - a_1^2)z], \tag{15c}$$

to determine the time evolution of the density matrix elements y_1, y_2, z. (The element x does not appear in (14).)

Under the conditions of weak pumping $(T/T_p \ll 1)$ and short pulses $(\omega_0 T \ll 1)$ one can neglect in the right-hand side of the above equations all terms involving the elements of ρ_e (i.e., y_1, y_2, and z) and assume $x = 1$. This is the linear absorption approximation discussed previously. The equations can be integrated immediately and lead to the density matrix (after pulse passage)

$$\rho_{e+} = \begin{pmatrix} y_{2+} & z_+ \\ z_+^* & y_{1+} \end{pmatrix} = k_0(\infty)\begin{pmatrix} a_2^2 & a_1 a_2 \\ a_1 a_2 & a_1^2 \end{pmatrix} \tag{16a}$$

with resulting quantum beat signal

$$I_0 = k_0(\infty)[a_1^2 b_1^2 + a_2^2 b_2^2 + 2a_1 a_2 b_1 b_2 \cos(\omega_0 t)]e^{-\Gamma t}, \tag{16b}$$

where the "preparation factor"

$$k_0(\infty) = \int_{-\infty}^{\infty} \frac{dt'}{T_p(t')} \tag{16c}$$

is a measure of the efficiency of excited state preparation. The beats occur at frequency $\omega_0/2\pi$ with a modulation depth η_0 (equivalent to the visibility of fringes in Young's two-slit experiment) given by

$$\eta_0 = 2\frac{a_1 a_2 b_1 b_2}{(a_1^2 b_1^2 + a_2^2 b_2^2)}. \tag{17}$$

Under conditions where $a_1 = a_2$ and $b_1 = b_2$ the modulation depth can reach 100%. This occurs, for example, in the case of "Zeeman quantum beats" following a $J = 0$ to $J = 1$ transition where both the excitation and decay radiation are polarized perpendicular to the external magnetic field, and only $m_J = \pm 1$ substates are excited.

Let us continue to assume that the pulse duration T is short compared to the quantum beat period $2\pi/\omega_0$, and that the effects of light shifts are negligible

$$\omega_0 T \ll 1, \qquad (\Delta E)T \ll 1, \tag{18}$$

but that the pumping need no longer be weak. One can therefore drop from (15b, c) those terms in which ω_0 and ΔE appear. Surprisingly, integration of the optical pumping equations gives rise to an excited state density matrix and quantum beat signal I of the same form as in (16a) and (16b), but with

$k_0(\infty)$ replaced by the time-dependent preparation factor

$$k(t) = \frac{1}{2(a_1^2 + a_2^2)}\left[1 - \exp\left(-\int_{-\infty}^{t} 2(a_1^2 + a_2^2)dt'/T_p(t')\right)\right]. \qquad (19)$$

If the pumping is weak, relation (19) reduces to (16b) upon a first-order expansion of the exponential. If the pumping is sufficiently strong, the exponential term becomes negligible, and $k(\infty)$ reduces to a constant and is independent of the pumping time $T_p(t)$. Regardless of the pumping strength, however, the signals I and I_0 are proportional. In other words, the strength of pumping, as usually expressed by the so-called saturation parameter

$$S \equiv T/T_p, \qquad (20)$$

has no influence on the modulation depth under the experimental condition of short pulse duration.

The physical significance of this result can be understood as follows. The evolution of the system may be thought of as a sequence of absorption and stimulated emission processes occurring on average every T_p seconds. In the weak-pumping approximation (with $S \ll 1$) one takes account only of the first absorption process which, starting from the ground state g, creates the excited-state populations y_1 and y_2 and the coherence z proportional to a_1^2, a_2^2, and $a_1 a_2$, respectively. For an intense pulse (with $S \gg 1$) many such sequences of absorption and stimulated emission can occur during the passage of one pulse. Nevertheless, stimulated emission destroys the excited state populations and coherence in the same proportions as absorption creates them with the consequence that the modulation depth is unaltered. Equation (14) shows that the modulation depth can change only if the relative magnitude of the excited state coherence and populations change.

If the system has a more complex level structure than that of a three-level atom, the form of the quantum beat signal can change with increasing saturation. Consider, for example, an atomic system with a pair of ground states, each one coupled by the laser pulse (and by spontaneous emission) to a distinct pair of excited states. Analysis of such a system leads to an optical signal that is the sum of two contributions of the form of (16b) with different preparation factor, excitation and decay matrix elements, and beat frequency for the two uncoupled sets of three states. In this case the ratio of the preparation factors and the modulation depth of each frequency component will depend on the saturation parameter T/T_p although, in general, this dependence is not very marked.

The effects of saturation show up in a much more interesting and surprising way when one considers the example of a long pulse excitation, i.e., a pulse length no longer negligible in comparison to the beat period ($\omega_0 T \gg 1$). Intuitively, one might anticipate that the contributions to the signal from each (differentially) small time interval during passage of the pulse would interfere destructively when spread over a total interval T that exceeds the period of beats to be observed. The modulated component of the optical signal should then diminish and eventually vanish as T is lengthened beyond $2\pi/\omega_0$. In the case of *weak* optical pumping, this expectation is indeed correct as will now be demonstrated.

Retaining ω_0 in (15c) for the coherence z—but neglecting the excited state populations and level shifts—leads to

$$z_+ = a_1 a_2 \int_{-\infty}^{\infty} \frac{\exp(-\omega_0 t')\, dt'}{T_p(t')}. \tag{21}$$

In the weak-pumping limit the populations y_{1+} and y_{2+} are still independent of ω_0 and given in the density matrix (16a). The quantum beat signal then takes the form

$$I_0(\omega_0) = [k_0(\infty)(a_1^2 b_1^2 + a_2^2 b_2^2) + 2k_{\omega_0}(\infty)\, a_1 a_2 b_1 b_2 \cos(\omega_0 t)] e^{-\Gamma t}, \tag{22a}$$

where

$$k_{\omega_0}(\infty) = \int_{-\infty}^{\infty} \frac{\exp(-i\omega_0 t')\, dt'}{T_p(t')} \tag{22b}$$

is the Fourier transform of the pulse profile at the beat frequency ω_0. For the time-independent or "dc" component of the beat signal, the preparation factor $k_0(\infty)$ is the Fourier transform at frequency 0. Compared to the corresponding case of a short pulse width, the modulation depth is now a function of the beat frequency

$$\eta = \eta_0 \frac{k_{\omega_0}(\infty)}{k_0(\infty)}. \tag{23}$$

From the integral in relation (22b) it is seen that the modulation depth—and hence the visibility of the beats—vanishes when the pulse duration is sufficiently long to permit variation of the phase $\omega_0 t'$ over at least 2π radians.

Intuition fails, however, in the case of both long pulse duration and *strong* pumping. If the linear approximation is no longer valid and the pulse duration no longer negligible, the entire set of coupled optical pumping equations must be solved as they stand. We will disregard for the moment, however, the light shifts and set $\Delta E = 0$. This is always possible if one restricts attention to a symmetric excitation profile centered on the frequency of the optical transition. Nevertheless, this simplification does not permit the equations to be solved in a simple, physically interpretable analytic form, and it is necessary to resort to numerical analysis by computer.

Let us adopt a Gaussian pulse profile

$$1/T_p(t) = \xi \exp\left(-\frac{t^2 \ln 2}{(T/2)^2}\right) \tag{24}$$

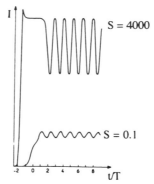

Figure 4.3. Restitution of quantum beats for excitations of sufficiently high intensity (as gauged by the saturation parameter S) and pulse length long with respect to the beat period. Under conditions of weak pumping (small S) a long pulse length leads, as shown, to low beat visibility. The theoretical parameters for the above curves are: $a_1 = a_2 = b_1 = b_2 = 1$; $\omega_0 T = 5$; $\Gamma = 0$. (Adapted from Silverman et al. [12].)

with an amplitude ξ which determines the saturation parameter

$$S = \int_{-\infty}^{\infty} dt / T_p(t). \tag{25}$$

In Figure 4.3 are shown computer simulations of the quantum beat signal for two pumping strengths—one low ($S = 0.1$) and one high ($S = 4000$). For purposes of illustration, the excitation and detection matrix elements have all been chosen to be unity, and the condition of long pulse duration is expressed by the assignment $\omega_0 T = 5$. Since it is the modulation depth, and not the exponential decay, that is of significance here, the beat profiles are shown for decay rate $\Gamma = 0$.

Although the modulation depth is seen to be small in the lower trace, as expected, where the pumping is weak, the striking feature of Figure 4.3 is that strong pumping gives rise to large beat visibility in the upper trace even though the pulse duration is long compared to the beat period. Why are the beats not "washed out" as in the case of weak pumping?

This effect, which has yet to be experimentally confirmed, can be understood qualitatively in terms of a random-walk model. It is useful to recall first that a coupled two-level quantum system (with periodic transitions between the two levels) may be likened to a classical electric dipole moment undergoing precession. (This point will be examined in greater detail in the next chapter when we consider the effects of resonant radiofrequency fields on excited states.) It is the off-diagonal elements—the coherence term z—in the density matrix that corresponds to the classical precession; the conjugate pair z and z^* represent precessional motion with opposite senses. We have previously described the time evolution of the atomic system as a succession of absorption and stimulated emission processes each occurring on average every T_p seconds. When the atom evolves freely in the excited manifold, the coherence z is said to precess at the angular frequency ω_0. However, this precession is eventually interrupted by a stimulated emission coupling z to the ground-state population x. (Since the state lifetime $\tau \gg T$, the effect of spontaneous emission can be neglected over the duration of the pulse.) When a subsequent absorption process occurs, it renews the excited states (both populations and coherence), but the precession of the coherence term can take place either in the original sense (x coupled to z) or in the opposite sense (x coupled to z^*).

During passage of the pulse the number of elementary absorption and emission processes is of the order of T/T_p. Over the interval of time ($\sim T_p$) that an atom temporarily remains in the excited manifold, the phase of the coherence term—i.e., the precession angle of the analogous classical dipole—is of the order $\omega_0 T_p$. It is assumed in this heuristic argument that the laser pulse is sufficiently intense, and the pumping time correspondingly short, that $\omega_0 T_p \ll 1$. Although the precession can occur sometimes in one direction, sometimes in the other, so that the mean precession angle is zero, the *dispersion* in phase is not zero, but is given by the root mean square of the individual phase variations

$$(\Delta\phi)^2 = (\omega_0 T_p)^2 \left(\frac{T}{T_p}\right) = \omega_0^2 T_p T. \tag{26a}$$

In order that the pulse create a substantial coherence in the excited state, the dispersion in phase must be less than about 1 radian ($\Delta\phi \leq 1$) which, from (26a) is equivalent to requiring

$$S = T/T_p \geq (\omega_0 T)^2. \tag{26b}$$

Thus, one learns from (26b) that even if the pulse duration is long ($\omega_0 T \gg 1$), as long as the saturation parameter—or, equivalently, the pumping strength—is sufficiently high, it should still be possible to observe quantum beats in the fluorescence signal.

An alternative way of considering this phenomenon is that the effect of saturation is to slow down the coherence precession rate—in the analogous way that virtual absorption and emission of light propagating through a transparent material lead to a phase velocity below c (or refractive index greater than unity)—and thereby prevent the destructive interference from wiping out the beats during passage of the pulse. Once the pulsed excitation is completed, the coherence term resumes its normal precession rate, and the modulation of the fluorescence occurs at the frequencies characterizing the energy level structure of the field-free atom. Although the saturation regeneration of quantum beats has not yet been observed because of the high value of S required ($S > 100$), the slackening of the precession rate, as a result of nonlinear interactions with the exciting light, has been reported in other types of optical pumping experiments, as, for example, the Hanle effect [23]. (The Hanle effect refers to the variation in fluorescent intensity of specified polarization when an external static magnetic field is varied about its null value at which point the radiating Zeeman substates become degenerate.)

Up to this point there has been no (or very small) light-induced displacement of energy levels. We consider now situations in which light shifts could have a potentially significant effect on the quantum beat signal. To isolate this effect from other consequences of saturation connected with pulse length, assume at first that the free precession of the atom during passage of the pulse is negligibly small ($\omega_0 T \ll 1$). It may seem plausible that, if the light shifts are large and different for each substate of the excited manifold, the dephasing of the atomic coherence during passage of the pulse would lead to disappearance of the quantum beats. This reasoning is deceptive, however. One can show rigorously that, regardless of the magnitude of the displacement—and even if $(\Delta E) T \gg 1$, the level displacement has no effect on the signal under the currently assumed experimental conditions. This result follows from integrating the optical pumping equations (15b, c) with $\omega_0 = 0$. The solutions y_1, y_2, z obtained for $t \to +\infty$ are all independent of ΔE.

An explanation of the above puzzling result may be found in the symmetry of the original density matrix equations (6a, b). It will be seen that the effective Hamiltonian characterizing light shifts in the excited states

$$\Delta E P_e \varepsilon \cdot \mathbf{D} P_g \varepsilon^* \cdot \mathbf{D} P_e,$$

has exactly the same structure as the product of operators

$$(1/T_p) P_e \varepsilon \cdot \mathbf{D} P_g \rho_g P_g \varepsilon^* \cdot \mathbf{D} P_e,$$

governing the preparation of the excited manifold *provided the ground state is isotropic* (i.e., ρ_g is proportional to P_g). In other words, absorption of a photon puts the atom into an eigenstate of the effective Hamiltonian governing level displacement, and therefore cannot create an atomic coherence among these eigenstates evolving in time at frequencies comparable to $\Delta E/\hbar$ during passage of the pulse even if $(\Delta E) T$ is large. The saturation effects tied to light shifts in this case are rigorously null.

If, however, the phase $\omega_0 T \geq 1$, the foregoing reasoning is no longer valid. It is then necessary to consider contributions of both the atomic Hamiltonian and the effective light-shift Hamiltonian to the evolution of the excited state during passage of the pulse. The quantization axis for energy eigenstates depends in that case upon the relative magnitudes of ω_0 and ΔE and, in addition, varies in time. Furthermore, if the atomic system is more complex than the one studied here and is characterized by a nondiagonal ground-state density matrix, then the pumping operator above need no longer have the same symmetry as the effective light-shift Hamiltonian. Under that circumstance, optical excitation could create coherence terms during the passage of the pulse even for $\omega_0 = 0$.

In general, the cases for which light shifts play a role in the determination of the quantum beat signal are complicated situations in which other effects of saturation are involved as well. The manifestation of these effects demands particular experimental conditions (such as light pulses with strong non-resonant spectral components) not likely to be realized in an actual quantum beat experiment.

In concluding this section we will consider briefly the effects of external fields on the quantum beat patterns. Although the introductory remarks of this chapter specifically pointed to the avoidance of time-varying external fields as one of the advantages of quantum beat spectroscopy, there are nevertheless circumstances under which the use of external static fields can be conceptually and practically helpful. Indeed, as will be illustrated in the following chapter, the interaction of a coherently prepared atom with time-varying fields as well can lead to some very interesting physics.

To determine the time evolution of an atomic system one must add to the field-free Hamiltonian H_0 the Hamiltonian of the appropriate external interaction H_{ex} whose specific form differs according to whether an electric or magnetic field is involved. In the case of coupling to a static magnetic field $\mathbf{B}_0$, the Zeeman Hamiltonian (in the absence of hyperfine structure) is given by

$$H_{ex} = -\mu_B \mathbf{B}_0 \cdot (\mathbf{L} + 2\mathbf{S}), \qquad (27a)$$

where μ_B is the Bohr magneton

$$\mu_B = |e|\hbar/2mc = 9.2732 \times 10^{-21} \text{ erg gauss}^{-1}, \qquad (27b)$$

and $\mathbf{L}$ and $\mathbf{S}$ are, respectively, the orbital and spin angular momentum operators (in units of $\hbar$). For coupling of a fine-structure state $|nLJm_J\rangle$ (where $\mathbf{J} = \mathbf{L} + \mathbf{S}$) to a static electric field $\mathbf{E}_0$ the effective Stark Hamiltonian can be written in the form

$$H_{ex} = \sum_{\{n'L'J'm'\}} \frac{\mathbf{E}_0 \cdot \boldsymbol{\mu}_E |n'L'J'm'\rangle\langle n'L'J'm'|\mathbf{E}_0 \cdot \boldsymbol{\mu}_E}{E_{nLJ} - E_{n'L'J'}}. \qquad (28)$$

The energy level structure, which determines the possible quantum beat frequencies, is then obtained by diagonalizing $H_0 + H_{ex}$. In general, the actual spectral composition of the quantum beats following excitation by one or more lasers depends on the polarization of the light and the relative strengths of the parameters governing the internal (e.g., fine structure) and external interactions. Even for hydrogenic systems like atomic Rydberg states, the calculation of the quantum beat profiles is by no means a trivial matter, and details will be left to the literature [13]. It is of conceptual interest, however, to illustrate the effects of an external magnetic field in the case of an atom whose internal Hamiltonian includes a spin-orbit coupling term

$$H_{so} = \hbar A \mathbf{L} \cdot \mathbf{S}, \qquad (29)$$

where A is a measure of the strength of this fine-structure interaction (in units of angular frequency since $\mathbf{L}$ and $\mathbf{S}$ are in units of $\hbar$).

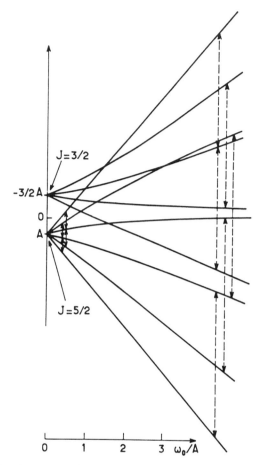

Figure 4.4. Variation in energy with magnetic field of the *nD* sublevels of the sodium atom. The fine structure ordering is reversed from that of hydrogen. Arrows indicate the expected transitions leading to quantum beats in the regions of low and high fields. (Adapted from Silverman et al. [13].)

Figure 4.4 shows the variation in energy of the *nD* sublevels of atomic sodium as a function of magnetic field strength. (Note that the ordering of the $J = \frac{3}{2}$ and $J = \frac{5}{2}$ levels is *opposite* that of normal ordering—i.e., the ordering of atomic hydrogen fine structure—as a result of complex inter-actions with the core electrons [24].) Since *D* and *S* states have the same parity, a direct transition from the 3*S* ground level cannot be effected by single laser excitation. Two laser pulses, however, properly timed and of appropriate frequency, can induce sequential 3*S*–3*P* and 3*P*–*nD* transitions thereby exciting the atom into a linear superposition of $D_{3/2}$ and $D_{5/2}$ states [25]. The relative orientation of the two excitation polarizations influences strongly the visibility of ensuing quantum beats; examination of a number of special cases suggests that greater beat contrast occurs for crossed polarizations.

In Figures 4.5, 4.6, and 4.7 are illustrated the quantum beat transients calculated numerically [13] for the respective conditions of zero, weak, and strong magnetic fields. The level of field strength is defined by the relative magnitude of the fine structure interaction parameter A and the cyclotron frequency

$$\omega_c = \mu_B B_0 / \hbar, \tag{30}$$

both of which enter the expression for the eigenvalues (in units of angular frequency)

$$\omega_{L\pm(1/2),m} = -\frac{A}{4} + m\omega_c \pm \tfrac{1}{2}[(Am + \omega_c)^2 + A^2(L(L+1) - (m+\tfrac{1}{2})(m-\tfrac{1}{2}))]^{1/2} \tag{31}$$

of the basis vectors $|Jm\rangle'$ of the Hamiltonian $H_0 + H_{ex}$. A configuration was chosen such that the two light polarizations are parallel to each other and perpendicular to the external magnetic field.

The theoretical zero-field quantum beat signal (Figure 4.5) displays a beat frequency $\omega_0 = 5A/2$, corresponding to the fine-structure level separation, with shallow modulation depth as expected. The application of a weak magnetic field ($\omega_0 \ll A$), however, enhances the beat visibility as shown in Figure 4.6. The signal is seen to be modulated primarily at the frequency $2g_J\mu_B B_0$ where $g_J = 1.2$ is the Landé factor of the $D_{5/2}$ level. The calculation shows that in this case the signal is particularly sensitive to the $\Delta m = 2$ coherence in the $D_{5/2}$ level. Contributions of other coherence terms either within the $D_{3/2}$ level or between the $D_{3/2}$ and $D_{5/2}$ levels evolving at different frequencies are much weaker. Arrows in the low-field region of the energy level diagram, Figure 4.4, indicate the couplings contributing to the quantum

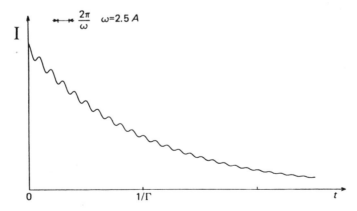

Figure 4.5. Theoretical zero-field quantum beat signal. (Adapted from Silverman et al. [13].)

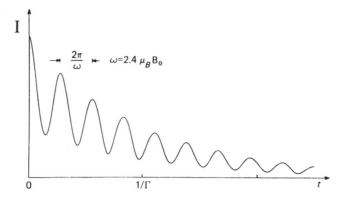

Figure 4.6. Theoretical weak field ($\omega_0 \ll A$) quantum beat signal. (Adapted from Silverman et al. [13].)

beat signal. Thus, measurement of the weak-field Zeeman beats can directly yield the Landé g-factor for excited states, a point of practical spectroscopic interest.

When the magnetic field is sufficiently strong ($\omega_0 \gg A$), the pattern of beats is completely changed, as Figure 4.7 shows. For the example of sodium D states the analytic form of the signal is

$$I = 198.15 + 105.55 \cos(At) \cos(\omega_L t), \tag{32}$$

where the two frequencies appearing in the signal are indicated by arrows on the high-field region of the energy level diagram, Figure 4.4. The signal now

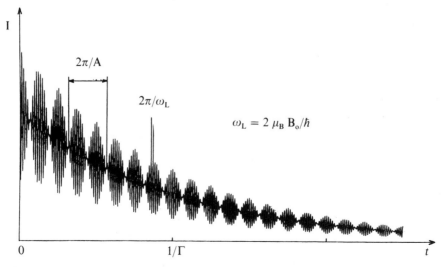

Figure 4.7. Theoretical strong field ($\omega_0 \gg A$) quantum beat signal. (Adapted from Silverman et al. [13].)

consists of a carrier wave at the Larmor frequency $\omega_L = 2\omega_c$ modulated strongly at the frequency A of the fine-structure interaction. What is the origin of this beats-within-beats structure?

A simple interpretation can be given in terms of the classical vector model of the atom [26]. In a high magnetic field the angular momenta $\mathbf{L}$ and $\mathbf{S}$ are decoupled and each can precess freely about the magnetic field. This explains the appearance of the Larmor frequency ω_L in place of $2g_J\omega_c$. One must also take account, however, of the diagonal part (Am_Lm_S) of the fine structure interaction, (29), which acts as a perturbation adding to the applied field a small internal magnetic field $B' = \hbar Am_S/\mu_B$. This internal magnetic field may be oriented parallel or antiparallel to the applied field according to the sign of $m_S = \pm\frac{1}{2}$, the projection of spin along the quantization axis (i.e., the direction of the external magnetic field). There are therefore two Larmor frequencies $\omega_{L\pm} = \omega_L \pm A$, and correspondingly the sum of two cosinusoidal terms at these frequencies gives the frequency dependence of relation (32)

$$\cos((\omega_L + A)t) + \cos((\omega_L - A)t) = 2\cos(At)\cos(\omega_L t).$$

The foregoing high-field quantum beam experiment, which has yet to be performed, suggests an alternative to zero-field experiments for measuring the fine structure interaction constant. As seen in the foregoing example, high-field quantum beats show a marked enhancement in the signal-to-noise ratio.

4.4. Correlated Beats from Entangled States

An entangled state, so designated by Schrödinger [27] in 1935, refers to a quantum state of a multi-particle system that cannot be expressed as a product of single-particle states. Such states give rise to correlations between separated particles which are unaccountable within the framework of classical physics and sometimes bizarre even by the expectations of quantum physics (if based on the study of single-particle systems or systems of uncorrelated particles). Entangled states are therefore of considerable interest to those concerned with the foundations of quantum physics. Schrödinger, himself, regarded the property of entanglement as the foremost characteristic property of quantum mechanics.

We have already encountered quantum entanglement in the discussion of the EPR paradox and the strange correlations manifested in the AB–EPR and AB–HBT experimental configurations of the previous chapter. It will be recalled that the EPR paradox was advanced as an argument that quantum mechanics could not be a complete theory and most likely had to be supplemented by additional (and possibly unknowable) variables. Since the appearance in 1935 of the seminal EPR paper [28], a number of experiments have been performed with entangled photon states for the purpose of revealing the possible existence of such local "hidden" variables by testing the correlations expressed in a set of inequalities derived by J.S. Bell [29]. An essential feature common to these EPR-type experiments is the correlation in space, time, or polarization of two photons emitted during radiative decay of an excited atomic state. In these experiments the two-photon entangled states arise via cascade transitions from single atoms excited incoherently by optical or electronic transitions [30]. More recently, the interferometry of photon pairs produced by the nonlinear optical process of parametric down-conversion has provided new examples of nonclassical light correlations [31].

In this section we examine a remarkable example of nonlocal correlations manifested by quantum beats in the radiative decay of entangled states of two identical, but widely separated, excited atoms [32]. What makes the example of particular interest, is that this quantum interference effect occurs even though each individual atom is *not* prepared in a linear superposition of excited states.

To appreciate the extraordinary nature of the effect recall that the spontaneous emission from incoherently populated atomic states decays exponentially in time. A percussional excitation sharply defined in time—produced, for example, by pulsed laser—and sufficiently uncertain in energy can put an atom into a linear superposition of close-lying excited states. Only then, it would seem, should the subsequent atomic fluorescence be modulated at the Bohr frequencies corresponding to the energy intervals of the superposed states. If the spectral width of the exciting pulse is much smaller than the Bohr frequencies, the atom will be prepared in a sharp energy eigenstate, and no quantum beats will be induced.

In striking contrast to the conditions characteristic of standard quantum beat spectroscopy, the quantum interference phenomenon to be described now can be produced by spectrally *narrow*, but correlated photons. The quantum beats then appear—not in the fluorescence from individual atoms—but in the *joint* detection of photon pairs, one light quantum arising from each of two separated atoms.

Figure 4.8 shows a schematic diagram of a possible experimental configuration. A light source—created, for example, by decay of S-state systems—produces photons in pairs, one photon of frequency ω_1 and the other of frequency ω_2, that propagate in opposite directions through polarizers ε_A and ε_B to excite two arbitrarily separated atoms. Which direction a photon of particular frequency takes is entirely random and unpredictable, although if it is known that an ω_1 has propagated to the left, then an ω_2 must necessarily have propagated to the right. Depending on whether the frequency is ω_1 or ω_2, a photon can induce a transition from the atomic ground state g to the excited state e_1 or e_2, respectively. Neither photon, however, prepares an atom in a linear superposition of states e_1 and e_2. Finally, the fluorescent photons created by radiative decay of the atoms from states e_1, e_2 to some final state f pass through polarizers ε_{DA} and ε_{DB} and are monitored without energy selection at detectors D_A and D_B.

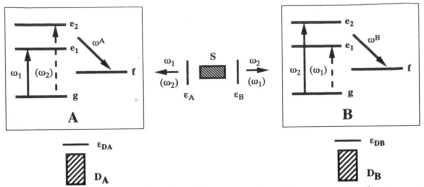

Figure 4.8. Configuration of a long-distance or EPR-type quantum-beat experiment. Pairs of photons emerge in opposite directions from source S exciting separated atoms A and B into state e_1 or state e_2, but *not* a superposition of the two. Photodetectors D_A and D_B individually reveal no quantum beats although beats appear at frequency $\omega_2 - \omega_1$ in their correlated outputs. (Adapted from Silverman [32].)

Time-ordered diagrams representing the two modes of excitation and decay are illustrated in Figure 4.9. Since the two modes give rise to indistinguishable final states of the whole system, their amplitudes are to be added, and the resulting probability for the process will display a quantum interference term. However, the *single-atom* density matrix, which reveals all that can be known by measurements on one of the two atoms (A or B), is equivalent (as will be demonstrated shortly) to that of an atom with *incoherently* populated excited states. Thus, no beats will appear at the output of one detector alone. The quantum beats occur only in the joint probability of receiving photons at the two detectors.

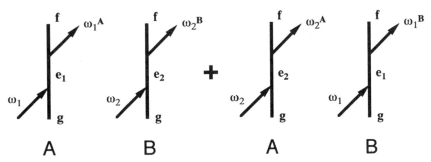

Figure 4.9. Time-ordered diagrams for the excitation and decay processes giving rise to long-distance quantum beats. Vertical lines represent the evolution of atoms A and B from ground state g to final state f via intermediate excited states e_1 or e_2. Oblique lines represent absorption (arrow in) and emission (arrow out) of photons.

Pursuant to the foregoing description, the state of the two-atom system immediately following excitation by two correlated photons is representable by the entangled superposition of state vectors

$$|\phi(0)\rangle = a_{12}|e_1^A; e_2^B\rangle + a_{21}|e_2^A; e_1^B\rangle, \tag{32a}$$

where the label e_j^K designates excited state j (1 or 2) of atom K (A or B). To within a constant factor unimportant for our present purposes the coefficients of the superposition can be written as

$$a_{ij} \sim \langle g|\boldsymbol{\mu}_A \cdot \boldsymbol{\varepsilon}_A|e_i\rangle\langle g|\boldsymbol{\mu}_B \cdot \boldsymbol{\varepsilon}_B|e_j\rangle, \tag{32b}$$

where $\boldsymbol{\mu}_K$ is the electric dipole moment of atom K. The density operator of the total system is then

$$\rho(0) = |\phi(0)\rangle\langle\phi(0)| \tag{33}$$

from which it readily follows by tracing over the states of atom B that the single-atom density operator of atom A takes the form

$$\rho_A(0) = \text{Tr}_B\{\rho(0)\} = |a_{12}|^2|e_1^A\rangle\langle e_1^A| + |a_{21}|^2|e_2^A\rangle\langle e_2^A|. \tag{34}$$

By tracing (33) over the states of atom A, one arrives at a similar expression for the single-atom density operator of atom B. Since the matrix represented by expression (34) is diagonal in an energy representation—i.e., displays only populations and no coherence terms—measurements made on A or B alone would not distinguish systems represented by state vector (32a) from two ensembles of atoms with incoherently populated excited states.

After the initial excitation, the states of the two atoms evolve freely and independently under their respective Hamiltonians. Thus, the state of atom A at time t_A and of atom B at time t_B (where the sharp excitation defines the time origin) is represented by the vector

$$|\phi(t_A, t_B)\rangle = a_{12}\exp\left[-i\left(\omega_1^A - \frac{i}{2}\Gamma_1^A\right)t_A\right]\exp\left[-i\left(\omega_2^B - \frac{i}{2}\Gamma_2^B\right)t_B\right]|e_1^A; e_2^B\rangle$$

$$+ a_{21}\exp\left[-i\left(\omega_2^A - \frac{i}{2}\Gamma_2^A\right)t_A\right]\exp\left[-i\left(\omega_1^B - \frac{i}{2}\Gamma_1^B\right)t_B\right]|e_2^A; e_1^B\rangle, \tag{35}$$

where ω_i^K is the angular frequency corresponding to excited state i of atom K, and Γ_i^K is the associated spontaneous emission decay rate. Although the two atoms A and B are of identical kind, the Bohr frequencies and decay rates are specifically labeled by A or B since the atoms may be moving with different velocities with respect to the stationary detectors of the laboratory frame and therefore subject to Doppler shifts.

The joint probability that one photon is received at detector D_A within an interval Δt_A about time t_A and a second is received at detector D_B within an interval Δt_B about time t_B is given by

$$P(t_A, t_B) = I(t_A, t_B)\,\Delta t_A\,\Delta t_B. \tag{36}$$

By generalization of the theory of quantum beats developed in the previous sections, one can determine $I(t_A, t_B)$ to within a constant instrumental factor by calculating the mean value of the detection operator

$$\mathcal{O}_{\text{det}} = [(\boldsymbol{\varepsilon}_{DA}\cdot\boldsymbol{\mu}_A)(\boldsymbol{\varepsilon}_{DB}\cdot\boldsymbol{\mu}_B)]|f^A; f^B\rangle\langle f^A; f^B|[(\boldsymbol{\varepsilon}_{DA}\cdot\boldsymbol{\mu}_A)(\boldsymbol{\varepsilon}_{DB}\cdot\boldsymbol{\mu}_B)]^\dagger \tag{37}$$

as follows

$$I(t_A, t_B) = \text{Tr}\{|\phi(t'_A, t'_B)\rangle\langle\phi(t'_A, t'_B)|\mathcal{O}_{\text{det}}\}, \tag{38a}$$

where

$$t'_K = t_K - r_K/c \tag{38b}$$

is the retarded emission time of atom K (A or B) a distance r_K from detector D_K.

Evaluation of the trace in (38a) leads to the joint intensity function

$$\begin{aligned}
I(t_A, t_B) &= |a_{12}b_{12}|^2 \exp[-(\Gamma_1^A t'_A + \Gamma_2^B t'_B)] + |a_{21}b_{21}|^2 \\
&\quad \times \exp[-(\Gamma_2^A t'_A + \Gamma_1^B t'_B)] \\
&\quad + 2\exp[-\tfrac{1}{2}((\Gamma_1^A + \Gamma_2^A)t'_A + (\Gamma_1^B + \Gamma_2^B)t'_B)] \\
&\quad \times \text{Re}\{a_{12}a_{21}^* b_{12}b_{21}^* \exp[-i(\omega_{21}^A t'_A - \omega_{21}^B t'_B)]\},
\end{aligned} \tag{39a}$$

in which

$$b_{ij} \sim \langle f|\boldsymbol{\mu}_A\cdot\boldsymbol{\varepsilon}_{DA}|e_i\rangle\langle f|\boldsymbol{\mu}_B\cdot\boldsymbol{\varepsilon}_{DB}|e_j\rangle \tag{39b}$$

are the spontaneous emission matrix elements, and

$$\omega_{21}^K = \omega_2^K - \omega_1^K \tag{40a}$$

is the beat frequency of the fluorescent emission from atom K in the laboratory frame. This frequency is related to the corresponding frequency in the atomic rest frame by the Doppler effect

$$\omega_{21}^K = \omega_{21}(1 + \mathbf{k}_K\cdot\mathbf{v}_K/c), \tag{40b}$$

where $\mathbf{k}_K$ is a unit vector from detector K to atom K and $\mathbf{v}_K$ is the velocity of atom K.

Since it is a Doppler-shifted *beat* frequency, rather than optical frequency, that comprises the signal, the quantum interference effect described here is largely insensitive to atomic motion. For example, for a thermal distribution of atoms at room temperature ($T \sim 300$ K), the atomic velocities span an approximate range $0 < v/c < 10^{-3}$, and the Doppler effect leads to a spread of $\sim 10^{11}$ s^{-1} about an optical frequency of $\omega_2 \sim \omega_1 \sim 10^{14}$ s^{-1}. This may well be much larger than the quantum beat frequency ω_{21} (derived, for example, from atomic fine structure) which could be typically of the order of 10^8 s^{-1} or lower. The Doppler spread of the beats, however, would for the above circumstances span the much smaller range $0 < \omega_{21} < 10^5$ and therefore have negligible effect on dephasing of the observed signal. Similarly, for an atom with sharply defined energy eigenvalues, $\omega_i \gg \omega_{21} \gg \Gamma_i \, (i = 1, 2)$, one can ignore the effect of atomic motion on the excited state decay rates.

Assume for puposes of illustration that the excitation and decay matrix elements expressed by relations (32b) and (39b) are real-valued numbers, and let us disregard the weak effects of atomic motion. Equation (39a) then reduces to the following simpler expression for the jointly detected two-photon fluorescent signal

$$I(t_A, t_B) = (a_{12}b_{12})^2 \exp[-(\Gamma_1 t'_A + \Gamma_2 t'_B)] + (a_{21}b_{21})^2 \exp[-(\Gamma_2 t'_A + \Gamma_1 t'_B)]$$
$$+ 2a_{12}a_{21}b_{12}b_{21} \exp[-\tfrac{1}{2}(\Gamma_1 + \Gamma_2)(t'_A + t'_B)]$$
$$\times \cos\{\omega_{21}[(t_B - t_A) - (r_B - r_A)/c]\}. \tag{41}$$

Note in particular the argument of the cosine factor of the interference term in which the retarded times have been replaced by the actual detection times by means of defining relation (38b). A significant feature of the above expression is that a large dispersion in the difference in retardation times can be tolerated without the quantum interference term being averaged away. Although in the foregoing discussion reference is made to "atom A" or "atom B," what is really implied, of course, are two distinct atomic ensembles within each of which many identical atoms are in motion and instantaneously located at different distances from the closest light detector. Thus, the spatial intervals r_A and r_B are distributed quantities. For the quantum beats to persist, the term $\omega_{21}|r_B - r_A|/c$ must remain less than about one radian as the optical path lengths from atoms A and B to all points on the detecting surfaces of D_A and D_B, respectively, vary. For a beat frequency ω_{21} of the order of 10^8 s^{-1}, one must have $|r_B - r_A| < c/\omega_{21} \sim 300 \text{ cm}$, which is experimentally easy to satisfy.

The possibility of quantum beats in the fluorescence of a multiatom system has long been known [20] for the traditional case of single-photon excitation of a localized system of atoms. In that case nearly contiguous atoms constitute a single quantum system, and quantum beats arise from the interference of decay pathways involving one or another of the associated atoms. These beats are highly sensitive to the dispersion in retardation times, as they are in general to the Doppler effect, and would be totally dephased for mean atom separations that exceed an optical wavelength ($\sim 10^{-5}$ cm). Moreover, such an interference can occur in elastic scattering only—i.e., the states g and f must be the same or the emitting atom can in principle be identified.

In striking contrast, the "long-distance" quantum beat phenomenon arising from entangled states can occur in inelastic ($f \neq g$) as well as elastic scattering, and is largely insensitive to atomic motion or location.

References

[1] E.B. Aleksandrov, Optical Manifestations of the Interference of Non-degenerate *Atomic States, Soviet Phys. Uspekhi*, **15**, 436 (1973) (quotation from p. 436).

[2] M.P. Silverman, On Measurable Distinctions Between Quantum Ensembles, *Ann. N.Y. Acad. Sci.*, **480**, 292 (1986).

[3] G. Breit, Quantum Theory of Dispersion (Continued). Parts VI and VII, *Rev. Mod. Phys.*, **2**, 91–140 (1933).

[4] P.A. Franken, Interference Effects in the Resonance Fluorescence of "Crossed" Excited Atomic States, *Phys. Rev.*, **121**, 508 (1961).

[5] E.B. Aleksandrov, Quantum Beats of Resonance Luminescence under Modulated Light Excitation, *Opt. Spectrosc. (USSR)*, **14**, 233–234 (1963); English translation of *Opt. Spektr.* **14**, 436–437 (1963).

[6] A. Corney and G.W. Series, Theory of Resonance Fluorescence Excited by Modulated or Pulsed Light, *Proc. Phys. Soc. (London)*, **83**, 207 (1964).

[7] R.A. Bernheim, *Optical Pumping: An Introduction* (Benjamin, New York, 1965).

[8] J.N. Dodd, W.J. Sandle, and D. Zisserman, Study of Resonance Fluorescence in Cadmium: Modulation Effects and Lifetime Measurements, *Proc. Phys. Soc.*, **92**, 497–504 (1967).

[9] E.B. Aleksandrov, Luminescence Beats Induced by Pulse Excitation of Coherent States, *Opt. Spectrosc. (USSR)*, **14**, 522–523 (1964); English translation of *Opt. Spektr.*, **14**, 957–958 (1964).

[10] E. Hadeishi and W. A. Nierenberg, Direct Observation of Quantum Beats Due to Coherent Excitation of Nondegenerate Excited States by Pulsed Electron Impact, *Phys. Rev. Lett.*, **14**, 891 (1965).

[11] I.A. Sellin, C.D. Moak, P.M. Griffins, and J.A. Biggerstaff, Periodic Intensity Fluctuations of Balmer Lines from Single-Foil Excited Fast Hydrogen Atoms, *Phys. Rev.*, **184**, 56 (1969).

[12] M.P. Silverman, S. Haroche, and M. Gross, General Theory of Laser-Induced Quantum Beats. I. Saturation Effects of Single Laser Excitation, *Phys. Rev. A*, **18**, 1507 (1978).

[13] M.P. Silverman, S. Haroche, and M. Gross, General Theory of Laser-Induced Quantum Beats. II. Sequential Laser Excitation; Effects of External Static Fields, *Phys. Rev. A*, **18**, 1517 (1978).

[14] G. Alber and P. Zoller, Laser Excitation of Electronic Wave Packets in Rydberg Atoms, *Phys. Rep.*, **199**, 231 (1991).

[15] M.P. Silverman, Anomalous Fine Structure in Sodium Rydberg States, *Amer. J. Phys.*, **43**, 244 (1980).

[16] C. Fabre, M. Gross, and S. Haroche, Determination by Quantum Beat Spectroscopy of Fine-Structure Intervals in a Series of Highly Excited Sodium D States, *Opt. Commun.*, **13**, 393 (1975).

[17] J.P. Barrat and C. Cohen-Tannoudji, Etude du Pompage Optique dans le Formalisme de la Matrice Densité, *J. Phys. Rad.*, **22**, 329 (1961).

[18] See, for example, C. Cohen-Tannoudji, Optical Pumping with Lasers, in *Atomic Physics*, vol. 4, edited by G. Zu Putlitz, E.W. Weber, and A. Winnacker (Plenum, New York, 1975), pp. 589–614.

[19] M.D. Crisp and E. T. Jaynes, Radiative Effects in Semiclassical Theory, *Phys. Rev.*, **179**, 1253 (1969).

[20] S. Haroche, Quantum Beats and Time-Resolved Fluorescence Spectroscopy, in *Topics in Applied Physics*, Vol. 13: *High-Resolution Laser Spectroscopy*, edited by K. Shimoda (Springer-Verlag, Heidelberg, 1976), p. 253.

[21] V. Weisskopf and E. Wigner, Berechnung der natürlichen Linienbreite auf Grund der Diracschen Lichttheorie (Calculation of the Natural Linewidth Based on the Dirac Theory of Light), *Z. Physik*, **63**, 54 (1930); Über die Natürliche Linienbreite in der Strahlung des harmonischen Oszillators (Concerning the Natural Linewidth in the Radiation of the Harmonic Oscillator), *ibid.*, **65**, 18. (1930).

[22] M.P. Silverman and F.M. Pipkin, Radiation Damping of Atomic States in the Presence of an External Time-Dependent Potential, *J. Phys. B: Atom. Molec. Phys.*, **5**, 2236 (1972).

[23] M. Ducloy, Nonlinear Effects in Optical Pumping of Atoms by a High-Intensity Multimode Gas Laser. General Theory, *Phys. Rev. A*, **8**, 1844 (1973).

[24] A qualitative discussion of this problem is given in M.P. Silverman, *And Yet It Moves: Strange Systems and Subtle Questions in Physics* (Cambridge University Press, New York, 1993). For a detailed relativistic treatment see E. Luc-Koenig, Doublet Inversions in Alkali–Metal Spectra: Relativistic and Correlation Effects, *Phys. Rev. A*, **13**, 2114 (1976).

[25] S. Haroche, M. Gross, and M.P. Silverman, Observation of Fine-Structure Quantum Beats Following Stepwise Excitation in Sodium D States, *Phys. Rev. Lett.*, **33**, 1063 (1974).

[26] See, for example, H.E. White, *Introduction to Atomic Spectra* (McGraw-Hill, New York, 1934), Chapter 10.

[27] E. Schrödinger, Discussion of Probability Relations between Separated Systems, *Proc. Cambridge Philos. Soc.*, **31**, 555 (1935).

[28] A. Einstein, B. Podolsky, and N. Rosen, Can Quantum Mechanical Description of Physical Reality Be Considered Complete?, *Phys. Rev.*, **47**, 777 (1935).

[29] J.S. Bell, On the Einstein–Podolsky–Rosen Paradox, *Physics*, **1**, 195 (1964).

[30] See Reference [4] of Chapter 3.

[31] Z.Y. Ou and L. Mandel, Violation of Bell's Inequality and Classical Probability in a Two-Photon Correlation Experiment, *Phys. Rev. Lett.*, **61**, 50 (1988).

[32] M.P. Silverman, Quantum Interference in the Atomic Fluorescence of Entangled Electron States, *Phys. Lett. A*, **149**, 413 (1990).

CHAPTER 5

Sympathetic Vibrations: The Atom in Resonant Fields

Go, rock the little wood-bird in his nest,
　Curl the still waters, bright with stars, and rouse
The wide old wood from his majestic rest,
　Summoning from the innumerable boughs
The strange, deep harmonies that haunt his breast

To The Evening Wind
William Cullen Bryant

5.1. Beams, Bottles, and Electric Resonance

Science, according to the articulate writer and Nobel Laureate in medicine, Peter Medawar, is the "art of the soluble" [1], and as there is virtually no system in nature that is *exactly* soluble, the application to the real world of physics—the quintessential science—is to a great extent the art of modeling. To recount the great successes of physics is in large measure to unfold an historical record of aptly chosen, albeit hypothetical, models of reality such as frictionless free-fall, point electrical charges, and ideal gases. Among these immensely useful abstractions is the two-level quantum system.

Quantum systems (depending on the nature of the potential energy function) ordinarily have many, perhaps an infinite, number of eigenstates, but can be regarded as having only two to the extent that one may neglect all couplings except those between a given pair of interest. Then the quantitative description of such a system can be cast into the same aesthetic mathematical form irrespective of whether the two states are the spin states of an electron, the hyperfine states of a hydrogen atom, the inversion modes of an ammonia molecule, or the macroscopic states of the two super-conducting regions comprising a Josephson junction. With an anticipated application to atomic physics in mind, we shall take the two-level system of this chapter to be an atom and consider its interaction with a classical linearly polarized monochromatic oscillating field. We have then the "purest" form of wave inducing transitions in the most elementary bound-state matter—and yet no exact analytical solution of this deceptively simple system has yet been found. How one can handle this situation will be addressed shortly.

The importance of the two-level atom to physics, however, can hardly be overestimated. Investigations of a two-level atom in an oscillating field go back at least to the mid-1930s, to the magnetic resonance experiments of I.I. Rabi whose interest in determining the signals of nuclear magnetic moments led him to derive the fundamental Rabi "flopping formula" [2] which has since served as a basis for nearly all successive magnetic resonance experiments. In this paper Rabi calculated the probability that an atom subject to a "gyrating" (or rotating) magnetic field of specified frequency (intended primarily in the radiofrequency domain) undergoes a transition from one to the other of its quantum states—or, viewed in terms of classical imagery, that the atomic magnetic moment undergoes a reorientation. The significance of this paper is succinctly captured by Rabi's biographer, J.S. Rigden: "Today, fifty years later, this paper is cited by laser physicists who use Rabi's "flopping formula," derived in the 1937 paper, thus showing how a great paper can be applicable far beyond the immediate intentions of its author" [3].

One particularly significant issue of Rabi's resonance method was the subsequent development of the atomic beam *electric* resonance technique pioneered by Lamb and Retherford [4] to make high-precision tests of relativistic atomic structure and quantum electrodynamics. Profiting from the development of radio- and microwave sources needed for radar during the Second World War, the Lamb–Retherford experiment marked the first major innovation in the study of excited-state atomic structure beyond traditional optical spectroscopy. The advantage of an atomic beam, as already mentioned, was that atoms could be studied in a region separate from the source of their production which was generally filled with rapidly fluctuating electromagnetic fields (as in the vicinity of an electrical discharge or an electron beam). Use of radiofrequency (rf) techniques permitted the direct coupling of excited states within the same electronic manifold with greatly reduced Doppler broadening.

In the Lamb–Retherford experiment, a beam of excited hydrogen atoms was produced by thermal dissociation of H_2 in a tungsten oven and then bombarded by a transverse beam of 10.8 eV electrons (whose energy corresponds to the transition $n = 1$ to $n = 2$). Except for the $2^2S_{1/2}$ metastable state, which has a relatively long lifetime of $\frac{1}{7}$ second due primarily to a two-photon transition to the $1^2S_{1/2}$ ground level (since single photon transitions are forbidden by angular momentum conservation), all the other excited states were rapidly extinguished. A virually pure beam of $2^2S_{1/2}$ atoms then passed through a microwave interaction region where $2^2S_{1/2}$–$2^2P_{1/2}$ electric dipole transitions were driven. This diminished the metastable population which ultimately reached the detector where they were observed by monitoring the electrons ejected by impact of the surviving atoms upon a tungsten plate.

The resonance curve or line shape, from which the $2^2S_{1/2}$–$2^2P_{1/2}$ level separation could be determined, is ideally a measure of the variation of transition probability with the frequency of the applied oscillating field. From a practical standpoint, however, it was not feasible at the time for the experimenters to sweep a microwave or rf oscillator without simultaneously encountering marked changes in the output power. Had they conducted the experiment under these conditions, the line shapes would have been distorted and therefore useless for rendering accurate information on atomic structure. Lamb circumvented this problem by keeping the oscillator frequency constant and varying, instead, the strength of a transverse, homogeneous magnetic field through which the atomic beam passed. The magnetic field served at least two purposes. First, by deflecting charged particles away from the detector, it reduced the noise background. Principally, however, by altering the separation of the magnetic substates within each fine structure level (the Zeeman effect), it afforded a means of selecting specific pairs of substates to investigate and of "sweeping out" the resonance line shape by scanning—not the applied frequency of the oscillator—but the resonance frequency of the atom.

The Lamb–Retherford experiment demonstrated conclusively the long-suspected failure of the Dirac relativistic theory of the hydrogen atom to predict correctly the energy separation of states of the same n, J quantum numbers. In addition, many fascinating points of atomic physics were brought to light. However, as a general method for investigating excited atomic states, this approach left much to be desired. For one, it was restricted to the study of $2S$ metastables, since only atoms thus prepared could survive traveling a macroscopic distance before decaying. The lifetime of a $2P$ state, for example, is approximately 2 ns; emerging from a 2500 K oven at a probable speed of nearly 8×10^5 cm/s, a $2P$ atom would not likely travel more than a few microns. There were also a large number of systematic effects, most traceable to the presence of the magnetic field, which distorted and/or shifted the resonance curves and required correction. Such effects included the variation of the transition matrix elements with magnetic field, the production (by the Lorentz transformation) of a "motional" electric field in the atom rest frame with resulting displacement of the line center (Stark effect) and selected quenching of hyperfine states, the unsymmetrical distribution of the hyperfine levels about the mean fine structure energy caused by incomplete decoupling of the nuclear and electron angular momenta (Back–Goudsmit effect), and "curvature" of the Zeeman levels due to partial decoupling of the electron orbital and spin angular momenta (Paschen–Back effect). In the words of Lamb [5]: "A lengthy programme of calculations and measurements is required to allow for all such sources of error."

To investigate non-metastable excited states, Lamb and his coworkers [6, 7] later introduced a new radio-frequency–optical technique in which the use of an atomic beam was abandoned for a resonance cell ("bottle" experiment). The method involved the excitation of hydrogen atoms to $3S$, $3P$, and $3D$ states by a low-energy electron beam, simultaneous irradiation by a rf field of fixed frequency, and detection of atomic transitions by monitoring the variation in Balmer α emission as a function of magnetic field strength. As is characteristic of a bottle experiment, production, irradiation, and detection all took place in the same region for which the "understanding of the origin and magnitudes of electric fields...was distinctly incomplete," in the words of Lamb and Sanders [6]. Electrical perturbations possibly due to space charge within the electron beam, charges on the walls of the glass vessel, motion of excited atoms across the magnetic field, and fluctuating fields of neighboring ions and electrons gave rise to line shifts and signal variations as conditions within the vessel changed with time. Since the sensitivity of hydrogen to electric perturbations increases rapidly with principal quantum number (the quadratic Stark shift, for example, varies roughly as n^6), the bottle method—in the period before quantum beat spectroscopy—promised serious difficulties as a general technique to probe excited atomic states.

If I have dwelt rather long upon Lamb's experimental methods, it is because—problems notwithstanding—they are the ingenious prototypes which, like Rabi's, inspired a variety of succeeding methods to expose the inner workings of the atom by approaching ever more closely the ideal of a two-level quantum system in a purely oscillating field. It is clear from the foregoing remarks that reaching such an ideal would be greatly facilitated if:

(i) atomic states were produced in quantity at one place and rapidly transported to a separate region for examination; and
(ii) state selection and spectroscopy were performed in the absence of an external static magnetic field.

One notably successful way this has been accomplished is by the electrostatic acceleration of atoms to a high (but nonrelativistic) speed and subsequent mapping out of resonance line shapes by the originally desired, and ideally simplest, procedure of sweeping the frequency ω of an applied rf or microwave field in an otherwise field-free environment. A representative experimental configuration is illustrated in Figure 5.1. Typically, the signal is obtained by counting photons in the decay radiation for a set period of time with the oscillating field off and on, respectively.

$$\text{Signal}(\omega) \equiv \frac{N(0) - N(\omega)}{N(0)}. \tag{1}$$

Since changes in the resonant response of the atoms with frequency modify the intensity and polarization of spontaneous emission and can therefore be detected optically, I have designated the method fast-beam optical electric resonance or OER [8]. The basic principles of OER spectroscopy, delineated in the early 1970s [9], still provide one of the most straightforward ways of probing atomic structure. A comparative summary of slow (thermal) beam, bottle, and OER methods is given in Table I.

Although the direct acceleration of neutral atoms to an energy of tens of

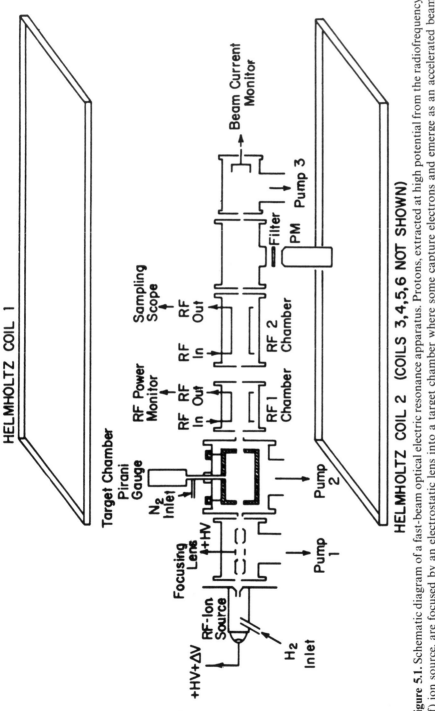

Figure 5.1. Schematic diagram of a fast-beam optical electric resonance apparatus. Protons, extracted at high potential from the radiofrequency (rf) ion source, are focused by an electrostatic lens into a target chamber where some capture electrons and emerge as an accelerated beam of neutral hydrogen atoms. The atoms pass through a spectroscopy chamber where desired transitions are induced by a rf field, then a state-selection chamber where undesired states are removed by a second rf field, and finally a detection chamber where decay photons are monitored by a photodetector (PM). The residual proton current is received in a Faraday cup. Three mutually orthogonal pairs of Helmholtz coils shield the apparatus from the Earth's magnetic field. (Adapted from C. Fabjan et al. [11].)

Table I. Comparison of spectroscopic methods.

	Slow beam	Bottle	Fast beam OER
Source	Thermal dissociation Electron bombardment	Electron bombardment	Charge-exchange conversion of a fast ion beam
Spectroscopy	Rf or static electric field Swept magnetic field	Rf electric field Swept magnetic field	Swept rf electric field Zero magnetic field
Detection	Metastable current	Decay radiation	Decay radiation
Limitations	Long-lived states Systematic errors from magnetic field	Complex electrical perturbations Variation in bottle environment	Specification of rf field

keV is not feasible, the conversion of accelerated protons by charge-capture collisions with gas or carbon foil targets [10] can provide a means of generating a beam of hydrogen atoms moving at speeds of the order of 10^8 cm/s, or nearly one hundredth the speed of light. As a consequence of the very short impact time, the fast-moving hydrogen atoms emerge from the target distributed over a large number of excited states with only a small loss in translational energy. The beam subsequently passes through one or more rf interaction regions in which transitions of interest are driven, while use of a precise power monitor ensures a constant rf power across the resonance line. Downstream from the rf interaction region the fluorescence of the atoms is detected as a function of the rf frequency. No external magnetic fields are applied, and the field of the Earth, itself, can be nulled by the use of Helmholtz coils.

An example of one of the first panoramic sweeps through the level structure of hydrogen is illustrated in Figure 5.2 for the $n = 4$ manifold [11]. The accompanying level diagram shows in broad outline the types of single-quantum electric dipole transitions that can occur. In the event that transitions between two different pairs of fine structure levels have neighboring resonance frequencies—as in the case of $4S_{1/2}-4P_{3/2}$ and $4P_{1/2}-4D_{3/2}$ transitions—the use of two or more separated rf interaction chambers can often prove helpful, one field serving as the "spectroscopy" field and the others as "quenching" fields to drive atoms in unwanted long-lived states into shorter-lived states that decay before reaching the detection chamber. Figure 5.2 illustrates state

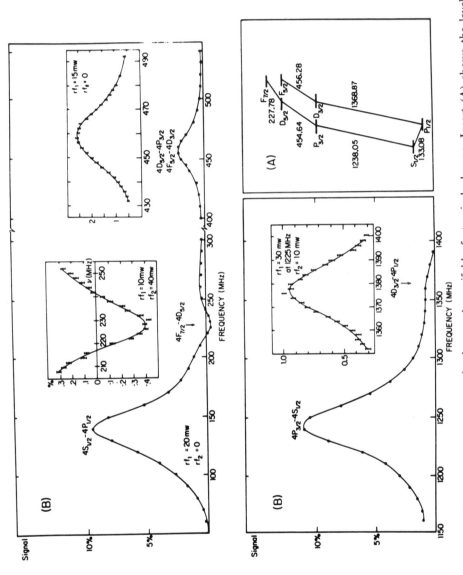

Figure 5.2. Panorama of fine structure resonances in the $n = 4$ manifold of atomic hydrogen. Insert (A) shows the level structure and frequencies of allowed transitions. Inserts in part (B) show line shapes resolved by rf quenching of overlapping transitions. (Adapted from C. Fabjan et al. [11].)

selection for the preceding pair of transitions. Generated by a single oscillating field, the $4P_{1/2}-4D_{3/2}$ resonance is seen as a poorly delineated plateau in the high-frequency tail of the $4S_{1/2}-4P_{3/2}$ resonance curve. However, with a quenching field set at 1225 MHz close to the resonance frequency of the $4S_{1/2}-4P_{3/2}$ transitions, the $4S_{1/2}$ states are eliminated, and the profile of the $4P_{1/2}-4D_{3/2}$ resonance curve shows up clearly.

Besides overlapping fine-structure resonances, the hydrogen spectrum is further complicated by the magnetic interaction between electron and nuclear spins which divides each fine structure level into two hyperfine components with total angular momentum quantum numbers $F = J \pm \frac{1}{2}$ and associated substates with magnetic quantum numbers: $-F \leq m_F \leq +F$. Thus, transitions at three different resonant frequencies—corresponding to "quantum jumps" with $\Delta F = \pm 1, 0$—can be induced between a pair of fine structure levels of opposite parity. In such a case sequential rf fields can be employed again to remove populations of atoms in unwanted hyperfine states. An example of hyperfine state selection is shown in Figure 5.3 where the $4P_{1/2}-4S_{1/2}$ resonance curve is simplified by total quenching of atoms in the $4S_{1/2}(F = 0)$ state.

In addition to furnishing resonance frequencies—and therefore details of the internal atomic structure—electric resonance spectroscopy is also sensitive to the initial relative populations of the coupled states, and consequently can provide significant information regarding the charge-changing interactions that created the atom. This information is coded in the exact shape of the resonance profile. Depending on the initial populations, the transitions induced by an oscillating field can lead to either a greater or lesser fluorescent emission than with the field turned off. As illustrated in Figure 5.2, most resonances are above the baseline; the oscillating field has driven the atoms from relatively long-lived states like $4S_{1/2}$ (lifetime 23 μs) to short-lived states like $4P_{1/2}$ (lifetime 12 ns) thereby diminishing the number of atoms that reach the detection chamber. The signal, (1), is then positive. However, transitions induced between initially populated states of comparable lifetime can suppress the rate of spontaneous decay, as in the case of coupled $4D_{5/2}$ (lifetime 36 ns) and $4F_{7/2}$ (lifetime 73 ns) states, giving rise to a resonance curve below the baseline. Since the net rate and direction of transitions between coupled states depend on the initial occupation probabilities, one would expect to see a variation in line shape with the choice of target atom from which an incoming proton captures an electron to form a neutral hydrogen atom. An example of this is illustrated in the upper row of Figure 5.4 for the overlapping transitions $D_{5/2}-P_{3/2}$ and $F_{5/2}-D_{3/2}$ of the $n = 4$ manifold. The lower row shows computer simulations of these spectra for different initial state assignments according to a spin-independent Coulomb interaction model of the electron capture process [8, 9].

Traditional charge transfer experiments employing fast particle detection are based on measurements of either the total current of ions produced in

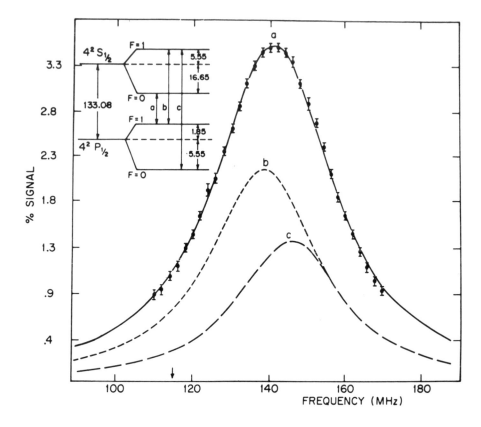

Figure 5.3. Example of hyperfine state selection. Hydrogen $4S_{1/2}-4P_{1/2}$ resonance with complete removal of atoms in the $4S_{1/2}$ ($F = 0$) hyperfine state and consequent suppression of transition a (as ascertained by a theoretical fit to the data of three Lorentzian line shapes). (Adapted from Silverman [8].)

various charge states, the attenuation of the primary beam, or the angular distribution of products scattered by violent close-encounter collisions. Such experiments yield information on the total electron capture cross sections [12, 13] and cannot distinguish scattering events which lead to different excited states of products of the same charge. Subsequent methods depending on electric field ionization are limited to manifolds of large principal quantum number and provide a sum over the cross sections of the component orbital sublevels. Similarly, methods that depend on the resolution of multistate radiative decay curves are incapable of resolving different states of the same lifetime such as fine or hyperfine structure states of the same orbital quantum number. By contrast, electric resonance line shapes differ in structure and occur in different portions of the frequency spectrum even for states of the same lifetime.

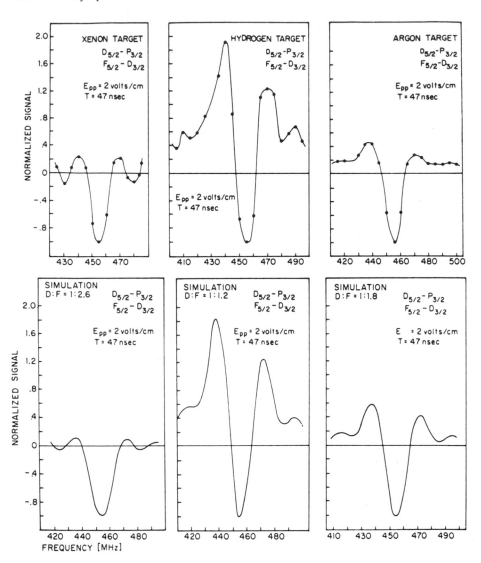

Figure 5.4. Sensitivity of electric resonance line shapes to initial state populations: example of hydrogen $4D_{3/2}$, $4D_{5/2}$, and $4F_{5/2}$ states prepared by electron capture collisions with different targets. Experimental spectra are shown above; theoretical spectra based on different initial D and F state amplitudes are shown below. (Adapted from Silverman and Pipkin [9], Part III.)

To extract both level structure and relative populations from an electric resonance line shape requires a detailed understanding of the interaction of a decaying multilevel atom with an applied oscillating field (and of the subsequent optical detection process as well). The analysis of so broad a problem goes well beyond the scope of this chapter and will be left to the original literature [8, 9]. It is grounded, however, in the study of the two-level atom in an oscillating field whose solution we will take up next, for, apart from spectroscopy, it is an important component of several experimental configurations exhibiting novel aspects of quantum interference.

5.2. Two Perspectives of the Two-Level Atom

The dynamics of the two-level atom with energy eigenstates $|1\rangle$, $|2\rangle$ is deducible from the Schrödinger equation which can be conveniently expressed in a matrix representation as follows:

$$H\Psi = i\,\partial\Psi/\partial t \quad \Rightarrow \quad \begin{pmatrix} h_{11} & h_{12} \\ h_{21} & h_{22} \end{pmatrix} \begin{pmatrix} c_1 \\ c_2 \end{pmatrix} = i\begin{pmatrix} \dot{c}_1 \\ \dot{c}_2 \end{pmatrix}, \tag{2a}$$

where

$$h_{ij} = \langle i|H|j\rangle, \qquad c_i = \langle i|\Psi\rangle \qquad (i, j = 1, 2). \tag{2b}$$

The constant $\hbar = h/2\pi$ is not shown explicitly, but has been divided out of both sides of the equation so that the elements of H are expressed in units of angular frequency. The diagonal elements of the Hamiltonian matrix (to be designated by the same symbol H as the Hamiltonian operator) characterize the energy of the states which can be complex numbers for decaying states where the real part is the actual energy, and the imaginary part is the decay rate; the off-diagonal elements govern transitions between the states. The state vector (in the terminology of Dirac) is quite literally a two-dimensional vector in this representation with elements—i.e., probability amplitudes—c_1 and c_2.

Any 2×2 matrix can be uniquely expressed as a linear superposition

$$H = h_0 1 + \sum_{i=1}^{3} h_i \sigma_i = h_0 1 + \mathbf{h} \cdot \boldsymbol{\sigma} \tag{3}$$

of the unit 2×2 matrix 1 and the Pauli matrices σ_i ($i = 1, 2, 3$) defined by the algebraic properties

$$\sigma_i \sigma_j = \sum_{k=1}^{3} \varepsilon_{ijk}\sigma_k + \delta_{ij}1, \tag{4a}$$

where ε_{ijk} is the completely antisymmetric tensor or Levi-Cività symbol, and δ_{ij} is the Kronecker delta symbol [14]. In the second equality of (3), $\boldsymbol{\sigma}$ is a three-dimensional vector whose elements are the Pauli matrices. There are infinitely many representations of the Pauli matrices (all related by unitary transformations) of which perhaps the most widely used is the following:

$$\sigma_1 = \begin{pmatrix} 0 & 1 \\ 1 & 0 \end{pmatrix}; \quad \sigma_2 = \begin{pmatrix} 0 & -i \\ i & 0 \end{pmatrix}; \quad \sigma_3 = \begin{pmatrix} 1 & 0 \\ 0 & -1 \end{pmatrix}. \tag{4b}$$

Adopting the above representation leads to the following elements of the Hamiltonian "four-vector" $H = (h_0, \mathbf{h})$ whose components are the coefficients of the basis vectors $1, \boldsymbol{\sigma}$:

$$h_0 = \tfrac{1}{2}(h_{11} + h_{22}); \quad h_1 = \tfrac{1}{2}(h_{12} + h_{21});$$

$$h_2 = \frac{1}{2i}(h_{21} - h_{12}); \quad h_3 = \tfrac{1}{2}(h_{11} - h_{22}). \tag{5}$$

If the Hamiltonian is Hermitian, $H^\dagger = H$, then $h_{21} = h_{12}^*$, and h_1, h_2 are, respectively, seen to be the real and imaginary parts of h_{21}.

Those familiar with physical optics will recognize the components h_0, $\mathbf{h}$ as an analogue to the Stokes parameters that uniquely characterize the polarization of a light beam [15]. The analogy follows because an arbitrarily polarized light wave can be described in terms of two mutually orthogonal basis states—for example, vertical and horizontal linear polarizations or left- and right-circular polarizations; light is consequently an example of a two-state system. (Since the quantum description of light is formulated in terms of photons, which are spin-1 bosons, one might wonder why there are only *two* components and not three, as is ordinarily the case for spin-1 particles. The answer, expounded in a delightful essay by Wigner [16], is intimately related to the fact that the photon has zero rest mass.)

In the above formalism the Schrödinger equation (2a) takes the succinct form

$$d\Psi/dt = -i(h_0 \mathbf{1} + \mathbf{h} \cdot \boldsymbol{\sigma})\Psi \tag{6a}$$

which can be integrated immediately

$$\Psi(t) = e^{-iHt}\Psi(0) = e^{-ih_0 t} e^{-i\mathbf{h}\cdot\boldsymbol{\sigma}t}\Psi(0), \tag{6b}$$

if the elements of H are independent of time. The first exponential factor in the second equality is an unimportant global phase factor which depends on the (arbitrarily adjustable) mean energy of the states and does not appear in expressions related to measurable quantities. (Note, however, that what may initially appear to be a global phase factor can have experimental consequences when the two coupled states must actually be regarded as a part of a larger quantum system. We will return to this point later in the chapter.) It is the second exponential factor of (6b) upon which the time-evolution of the system depends significantly. Operationally, expressions of the form of (6b) that involve exponential powers of matrices are defined by a Taylor series expansion of the exponential function

$$e^{-i\mathbf{h}\cdot\boldsymbol{\sigma}t} \equiv \sum_{n=0}^{\infty} \frac{(-i\mathbf{h}\cdot\boldsymbol{\sigma}t)^n}{n!} = 1\cos(ht) - i\frac{\mathbf{h}\cdot\boldsymbol{\sigma}}{h}\sin(ht), \qquad (7a)$$

in which the scalar h (not to be confused with Planck's constant which shall rarely be needed in this chapter) is the magnitude of the vector $\mathbf{h}$

$$h = \sqrt{\mathbf{h}\cdot\mathbf{h}} = \sqrt{h_1^2 + h_2^2 + h_3^2} = \tfrac{1}{2}\sqrt{(h_{11} - h_{22})^2 + 4h_{12}h_{21}}. \qquad (7b)$$

The reduction to the closed-form expression of (7a) follows straightforwardly from the algebraic properties of the Pauli matrices summarized in (4a) and from the Taylor series for sine and cosine.

One can also evaluate expressions of the form of (6b) by a more general procedure, the solution of an eigenvalue problem, that is independent of the dimension of the representation of H and therefore does not rely on the algebra of the Pauli matrices. It is instructive to examine this alternative method for it provides another mathematical interpretation of h_0 and $\mathbf{h}$. The idea is to diagonalize H—i.e., to transform it to a matrix η containing elements only along the principal diagonal: $\eta_{ij} = \eta_i\delta_{ij}$—since the exponential of a diagonal matrix $\exp(\eta)$ is readily shown to take the form of a diagonal matrix whose elements are $\exp(\eta_i)$. (To show this one employs again the Taylor series representation of the exponential function.) Then the solution given by the first equality of (6b) can be explicitly evaluated. By standard matrix techniques [17] one diagonalizes the matrix H by first solving the eigenvalue problem

$$HX_i = \eta_i X_i \qquad (8a)$$

which yields n eigenvalues η_i for an n-dimensional matrix, and relating H and η by the transformation

$$H = D\eta D^{-1}, \qquad (8b)$$

where η displays the eigenvalues of H along the principal diagonal in some designated order $\eta_1, \eta_2, \ldots, \eta_n$; the diagonalizing matrix D is then constructed by juxtaposing the corresponding eigenvectors X_i of H in the same order (from left to right)

$$D = X_1 X_2, \ldots, X_n. \qquad (8c)$$

By employing the Taylor series definition of the exponential yet again, one can demonstrate that

$$e^{-iHt} = e^{-iD\eta D^{-1}t} = De^{-i\eta t}D^{-1}, \qquad (9)$$

from which the solution to the Schrödinger equation in the case of a two-level system can be explicitly written

$$\Psi(t) = e^{-iHt}\Psi(0) = D\begin{pmatrix} \exp(-i\eta_1 t) & 0 \\ 0 & \exp(-i\eta_2 t) \end{pmatrix} D^{-1}\Psi(0). \qquad (10)$$

The equivalence of solutions (6b) and (10) is apparent once the eigenvalue problem is actually solved. In the case of the two-level atom, the two eigenvalues of H deducible from solution of a quadratic secular equation are

$$\eta_{1,2} = \tfrac{1}{2}[(h_{11} + h_{22}) \pm \sqrt{(h_{11} - h_{22})^2 + 4h_{12}h_{21}}]. \tag{11a}$$

The components of the "four-vector" $(h_0, \mathbf{h})$ are therefore related to the eigenvalues of H by

$$\eta_{1,2} = h_0 \pm h. \tag{11b}$$

If the elements of H depend on time, then the resulting solution does not, in general, take the simple form of relations (6b). Indeed, as pointed out at the beginning of the chapter, there may be no known exact analytical solution at all even in the "simple" case of a two-level atom with a harmonically oscillating field. Where a closed-form analytical solution is not achievable, one must then either seek an approximate solution to the original equations or simplify the model further to obtain an analytical solution to reduced equations. In the latter case the strategy is to remove from H by means of appropriate transformations as much of the time-dependence as possible, so that the solution to the transformed Schrödinger equation can be cast in the form of (6b). One way this has been traditionally done is to assume (although it is ordinarily not the case experimentally) that the atom is interacting with a *rotating* field, rather than with an *oscillating* field. In the next section we will discuss this procedure and a more general one, designated the "oscillating field theory" [18], which produces a closed-form analytical solution almost identical to the exact solution obtainable by numerical integration of the Schrödinger equation.

Before examining the specific problem of an external rf field, it is conceptually useful to re-examine the dynamics of a two-level system from the perspective of a second general mathematical formalism, that of the density operator ρ whose elements in a matrix representation enter directly into theoretical expressions for observable quantities. Consider first a system with two coupled *stable* states for which the Hamiltonian matrix is necessarily Hermitian. By multiplying the two sides of the Schrödinger equation (2a) by Ψ on the right, and subtracting from the result the Hermitian conjugate of the Schrödinger equation multipled by $\Psi^\dagger$ on the left, one obtains the equation of motion in the Heisenberg form

$$d\rho/dt = -i[H, \rho] \tag{12a}$$

(which we used, but did not derive, in the discussion of laser-induced quantum beats). Employing again the decomposition in (3), one can express the two-dimensional matrix representation of the density operator as

$$\rho \equiv \Psi\Psi^\dagger = \begin{pmatrix} \rho_{11}\rho_{12} \\ \rho_{21}\rho_{22} \end{pmatrix} = \rho_0 1 + \boldsymbol{\rho}\cdot\boldsymbol{\sigma}, \tag{12b}$$

where the expressions of ρ_0 and ρ in terms of the elements ρ_{ij} are formally the same as those given by (4) for the elements of H. The diagonal elements represent the relative populations of each level, and hence the trace $\mathrm{Tr}(\rho) = \rho_{11} + \rho_{22}$, to which ρ_0 is proportional, is a measure of the total probability of finding the atom in one or the other of the two available states. The off-diagonal elements characterize the coherence properties of the system, i.e., the capacity to produce quantum interference effects. (Note that in the definition of the density operator the order of the state vector and its Hermitian conjugate is significant; the reverse order $\Psi^\dagger\Psi$ produces a number, $\mathrm{Tr}(\rho)$, and not a matrix.)

Substitution of the expansion of ρ in (12b) into the commutator of (12a), use of the commutation properties of the Pauli matrices

$$[\sigma_i, \sigma_j] = 2i\varepsilon_{ijk}\sigma_k \tag{13}$$

that follow from (4a), and extraction of the coefficients of the bases 1 and σ lead to the scalar and vector equations of motion

$$d\rho_0/dt = 0, \tag{14a}$$

$$d\rho/dt = 2(\mathbf{h} \times \rho) \equiv \Omega_r \times \rho. \tag{14b}$$

In view of the interpretation of ρ_0, the first relation, (14a), is a statement of the conservation of probability for a closed system. If the two coupled states of interest were not stable, and the atom could decay radiatively to lower states whose dynamics were not of present concern, then ρ_0 would not be constant in time. We will see shortly how this case can be treated. The dynamics of the stable two-state system is effectively contained in relation (14b) which formally resembles the equation of motion of an angular momentum vector precessing at an angular velocity $\Omega_r = 2\mathbf{h}$. The relationship is only a formal one—nothing in the two-level atom need actually be precessing—but it provides a useful classical picture of the time evolution of the system.

The connection to rotation may also be seen directly in the solution for the state vector, (6b), since the unitary operator $U(\theta)$ for rotating a spin-$\frac{1}{2}$ system with angular momentum $\mathbf{S} = \frac{1}{2}\sigma$ (in units of $\hbar$) by an angle θ about the unit normal vector $\mathbf{n}$ is

$$U(\theta) = \exp(-i\sigma\cdot\mathbf{n}\theta/2). \tag{15a}$$

Comparison of (6b) and (15a) shows that the rotation axis is

$$\mathbf{n} = \mathbf{h}/h \tag{15b}$$

with angle of rotation

$$\theta = 2ht = \Omega_r t. \tag{15c}$$

Thus, one may imagine an ensemble of precessing dipoles as an analogue of the transitions between two quantum states. When the atoms are subjected to the interaction **h** over a time t, the corresponding dipoles process through an angle $\theta = 2ht$.

If the elements of H are independent of time, it follows from relations (6b) and (12b) that the time-evolution of the density matrix takes the form

$$\rho(t) = e^{-i\mathbf{h}\cdot\boldsymbol{\sigma}t}\rho(0)e^{i\mathbf{h}\cdot\boldsymbol{\sigma}t}. \tag{16}$$

Upon use of (7a) to evaluate (16) explicitly, one finds that the atomic system returns to its initial state described by $\rho(0)$ when the interaction is of such strength and duration that the precession angle θ equals π radians. One might have expected that the periodicity of θ should be 2π radians, but this is not the case for a two-level system. Although $\rho(\theta = \pi) = \rho(0)$, the corresponding state vectors differ by a sign: $\Psi(\theta = \pi) = -\Psi(0)$. As mentioned previously, global phase factors such as $e^{i\pi} = -1$ ordinarily have no observable consequences. However, if the two coupled states should subsequently be regarded as part of a larger system of states, then the once global phase may actually become a relative phase that can be manifested through a quantum interference experiment [19]. We shall examine this possibility in more detail in a subsequent section.

In analyzing the two-level atom—or, indeed, a quantum system with arbitrary number of levels— it is often convenient to remove at the outset the highest frequency terms in the diagonal elements of H corresponding to the energy eigenvalues. The precession of the system vector ρ then takes place more slowly under the influence of the weaker interactions, both internal and external, which are of principal interest. (It is assumed that the unperturbed energies have already been determined by solution of the appropriate eigenvalue problem, and that the evolution of the system under some time-dependent potential is of primary concern.) This simplification is effected by decomposing the Hamiltonian into a diagonal part H_0 yielding the unperturbed energies

$$H_0 = \begin{pmatrix} \omega_1 & 0 \\ 0 & \omega_2 \end{pmatrix} \tag{17a}$$

and an interaction V

$$V = \begin{pmatrix} v_{11} & v_{12} \\ v_{21} & v_{22} \end{pmatrix}, \tag{17b}$$

and then transforming to the so-called "interaction representation" by substituting

$$\Psi = e^{-iH_0 t}\Psi_I \tag{18a}$$

into the Schrödinger equation to obtain the transformed equation

$$H_I\Psi_I = i\,\partial\Psi_I/\partial t \tag{18b}$$

with interaction Hamiltonian

$$H_I = U_I^\dagger H U_I - iU_I^\dagger\,\partial U_I/\partial t = U_I^\dagger V U_I. \tag{18c}$$

If V does not contribute to the level energies—i.e., if V serves only to couple different states—then the resulting interaction Hamiltonian

$$H_I = \begin{pmatrix} 0 & v_{12}\exp(i\omega_0 t) \\ v_{21}\exp(-i\omega_0 t) & 0 \end{pmatrix} \tag{18d}$$

with energy interval (in frequency units)

$$\omega_0 \equiv \omega_1 - \omega_2, \tag{18e}$$

has no elements along the principal diagonal.

Under the preceding conditions where $v_{11} = v_{22} = 0$, the interaction vector V—in the four-vector representation $V = (V_0, \mathbf{V})$—has a null component V_3. If, in addition, $\mathbf{V}$ is purely real-valued, then $v_{12} = v_{21}$, and component $V_2 = \mathrm{Im}(v_{21})$ also vanishes. The angular velocity $\mathbf{\Omega}_r = 2\mathbf{h}$ then has components $\Omega_1 = 2v_{12}$ and $\Omega_3 = \omega_0$. For transitions induced by an optical interaction between the atomic ground state and an electronic excited state, the frequency ω_0—if atomic energy levels are to have any meaning—must be much greater than the corresponding frequency $2v_{12}$ of the interaction. In that case, one could regard the system vector as precessing around the "3-axis" at the rate ω_0 with a much slower nutational motion about the "1-axis" at the frequency $2v_{12}$. However, for the anticipated example of rf-coupling of atomic fine or hyperfine states, the frequencies ω_0 and $|\mathbf{V}|$ can be of comparable magnitude. Let us examine this point quantitatively.

The energy of an electron with principal quantum number n in a hydrogenic atom with atomic number Z

$$E_n \sim \frac{Z^2}{n^2} Ry \tag{19a}$$

is proportional to the Rydberg unit, $Ry \equiv e^2/2a_0$, where e is the electron charge and a_0 is the Bohr radius. The splitting of this level arising from the interaction of the electron spin and orbital motion (fine structure) is smaller by the square of the fine structure constant $\alpha_{fs} = e^2/\hbar c \sim 1/137.07$, or more precisely

$$\omega_0 \sim \frac{Z^2 \alpha_{fs}^2}{n} E_n \sim \frac{Z^4 \alpha_{fs}^2}{n^3} Ry. \tag{19b}$$

Since the orbital radius $r(n, Z)$ of the electron varies in the Bohr model as n^2/Z, the electric dipole interaction between the electron and an external oscillating electric field of amplitude E_0 can be roughly estimated by

$$v_{12} = er(n, Z)E_0 \sim \frac{en^2 a_0}{Z} E_0. \tag{20}$$

Comparing expressions (19b) and (20) one has

$$\frac{v_{12}}{\omega_0} \sim \frac{n^5 E_0}{\alpha_{fs}^2 Z^5 E_a},$$ (21a)

where the atomic unit of electric field strength is

$$E_a = e/a_0^2 = 5.14 \times 10^9 \text{ V/cm}.$$ (21b)

Thus, in the case of the hydrogen atom ($Z = 1$) subjected to an external field of $E_0 \sim 10$ V/cm for example, the ratio of precession frequencies v_{12}/ω_0 would be about 10^{-3} in the level $n = 2$ and of nearly equal magnitude in $n = 8$. The fact that the interaction with an external rf field can be comparable to (or even exceed) the internal interactions that split the coupled levels makes the analysis of the rf resonance problem in some ways more difficult than that of optical resonance for which the ratio v_{12}/ω_0 is ordinarily very small.

We conclude this section by taking into account the radiative decay of the coupled states, for it is by atomic fluorescence that the rf-induced transitions are observed in optical electric resonance experiments. As discussed previously, unstable levels can be characterized by a complex frequency of the form $\omega - i\gamma$ where $\hbar\omega$ is the energy eigenvalue of the state in the absence of spontaneous emission and γ is the decay rate, or reciprocal of the mean lifetime τ. One then adds to the Hamiltonians (17a, b) for stable levels a pure imaginary diagonal decay operator whose representation as a 2×2 matrix takes the form

$$-i\Gamma = -i\begin{pmatrix} \gamma_1/2 & 0 \\ 0 & \gamma_2/2 \end{pmatrix} = -i(\Gamma_0 1 + \boldsymbol{\Gamma} \cdot \boldsymbol{\sigma}).$$ (22)

The total Hamiltonian is now no longer Hermitian, and the procedure by which the Heisenberg form of the equation of motion (12a) was derived now leads to an equation of the form

$$d\rho/dt = -i[H, \rho] - \{\Gamma, \rho\},$$ (23a)

where the braces again signify the "anticommutator," $\{A, B\} \equiv AB + BA$. (See (4.6a, b). Note that the elements of Γ_e differ from those of Γ in (22) by a factor of 2.) The components of the density matrix (in the 1, $\boldsymbol{\sigma}$ basis) then satisfy the coupled relations

$$d\rho_0/dt = -2(\Gamma_0\rho_0 + \boldsymbol{\Gamma} \cdot \boldsymbol{\rho}),$$ (23b)

$$d\boldsymbol{\rho}/dt = 2(\mathbf{h} \times \boldsymbol{\rho} - \Gamma_0\boldsymbol{\rho} - \rho_0\boldsymbol{\Gamma}),$$ (23c)

where the vector $\mathbf{h}$ comes from the Hermitian part of the Hamiltonian ($H_0 + V$). From (23b) it is seen that the probability of finding the atom in one or the other of the two specified levels is no longer conserved, but decays exponentially. The level instability also influences the precession of the system vector $\boldsymbol{\rho}$ in accordance with (23c).

5.3. Oscillating Field Theory

The foregoing equations of motion of a two-level decaying atom, whether expressed in terms of amplitudes or density matrix elements, have no known exact analytical solution for the case of a pure oscillatory interaction. To the extent that high accuracy is not required in the study of line shapes resulting from the associated resonance experiments, an exact solution can be obtained for a simplified interaction, that of the rotating field (or rotating wave) approximation, first employed by Rabi in his derivation of the "flopping formula." The underlying physical idea, expressed in classical imagery, is that a linear oscillation can be decomposed into a sum of two counterrotating motions, one of which will be nearly resonant with the precessing dipole moment (representative of the quantum transitions between two states), and the other antiresonant, i.e., rotating in a sense opposite that of the dipole precession. From a classical perspective the component of the applied field rotating with the dipole exerts a steady torque whose effect is cumulative over many precession periods. The torque exerted by the counterrotating component, however, reverses itself at a rate 2ω and therefore might be expected to have no significant long-time effect. The rotating wave approximation consists of neglecting all terms at the counter-rotating frequency.

Unless, of course, one actually has a rotating field, neglect of the antiresonant interaction has theoretical consequences. The first consideration of these effects was given by Bloch and Siegert [20] and by Stevenson [21] who showed that the resonance maximum undergoes a small displacement. These results, derived for the case of magnetic resonance in a stable spin-$\frac{1}{2}$ system, are not *a priori* applicable in the case of decaying states. In this section the problem of a two-level atom subjected to an oscillating field is examined more generally [18] by a procedure I have designated the oscillating-field (or oscillating-wave) approximation. By means of several matrix transformations a representation is found in which the major effects of the applied field are independent of time—so that the procedure of the foregoing section is applicable—and the residual time-dependent portion contributes negligibly to observable atom-field interactions. It will be seen that, besides a frequency shift, the linearly oscillating field alters as well the difference in level decay rates, and indeed can affect the details of the overall line shape. The oscillating field approximation yields results which are virtually identical to those obtained by exact numerical integration of the Schrödinger equation.

Our quantum system consists of two "quasi-stationary" states $|1\rangle$, $|2\rangle$ interacting through the electric dipole moment μ_E with a classical oscillating field $\mathbf{E}_0 \cos(\omega t)$ of well-defined phase (here set to be zero). These atomic states are quasi-stationary in the sense that, although they are energy eigenfunctions of a Hermitian Hamiltonian, their interaction with the vacuum electromagnetic field results in a finite lifetime and hence level width.

For well-defined states to exist, however, the level width must be much smaller than the energy eigenvalue, or $\omega_i \gg \gamma_i$ (for $i = 1, 2$). The dynamics of the atom–rf field system is described by the Hamiltonian operator

$$H = H_0 - i\Gamma - \boldsymbol{\mu}_E \cdot \mathbf{E}_0 \cos(\omega t) \tag{24a}$$

which leads to a matrix differential equation of the form of (2a)

$$\begin{pmatrix} -(i\omega_1 + \tfrac{1}{2}\gamma_1) & -iV_{12}(e^{i\omega t} + e^{-i\omega t}) \\ iV_{12}(e^{i\omega t} + e^{-i\omega t}) & -(i\omega_2 + \tfrac{1}{2}\gamma_2) \end{pmatrix}\begin{pmatrix} c_1(t) \\ c_2(t) \end{pmatrix} = \frac{d}{dt}\begin{pmatrix} c_1(t) \\ c_2(t) \end{pmatrix}, \tag{24b}$$

in which

$$V_{12} = \frac{\langle 1|-\boldsymbol{\mu}_E \cdot \mathbf{E}_0|2\rangle}{2} \tag{24c}$$

is the interaction matrix element in units of frequency.

Upon transforming to the interaction representation, as described in the previous section (18a–c), one eliminates from (24b) the high-frequency eigenvalues of H_0 to obtain the equation

$$\begin{pmatrix} -\gamma_1/2 & -iV_{12}(e^{i\tilde{\Omega}t} + e^{-i\Omega t}) \\ -iV_{12}(e^{-i\tilde{\Omega}t} + e^{i\Omega t}) & -\gamma_2/2 \end{pmatrix}\Psi_1 = \frac{d}{dt}\Psi_1, \tag{25}$$

where

$$\Omega = \omega - \omega_0 \tag{26a}$$

is the deviation of the applied frequency from the unperturbed resonance frequency ω_0, and

$$\tilde{\Omega} = \omega + \omega_0 \tag{26b}$$

is the frequency of the antiresonant component.

At this point the rotating-field approximation is usually invoked by neglecting all terms containing $\tilde{\Omega}$. It is of conceptual interest to complete the rotating-field solution for decaying states before taking up the more accurate oscillating-field solution. The problem is in fact solved by (6b) once a transformation is found to remove the residual resonant time dependence from the matrix of (25). This task is accomplished by the unitary transformation

$$\Psi_R = \begin{pmatrix} e^{i\Omega t/2} & 0 \\ 0 & e^{-i\Omega t/2} \end{pmatrix}\Psi_1 \tag{27a}$$

which leads to the time-independent Schrödinger equation in the rotating frame

$$\begin{pmatrix} -(\gamma_1 - i\Omega)/2 & -iV_{12} \\ -iV_{12} & -(\gamma_2 + i\Omega)/2 \end{pmatrix}\Psi_R = \frac{d\Psi_R}{dt}. \tag{27b}$$

Upon integrating (27b) and making the subsequent inverse transformations back to the original representation of (24b), one obtains the complete solution

$$\Psi(t) = \begin{pmatrix} \exp(-(\gamma_1/2 + i\omega_1)t) & 0 \\ 0 & \exp(-(\gamma_2/2 + i\omega_2)t) \end{pmatrix} \begin{pmatrix} I_{11} & I_{12} \\ I_{21} & I_{22} \end{pmatrix} \Psi(0), \quad (28)$$

where the elements I_{ij} of the interaction matrix are

$$I_{11} = \exp\{(G - i\Omega)t/2\}\left[\cos(vt) - \left(\frac{G - i\Omega}{2v}\right)\sin(vt)\right], \quad (29a)$$

$$I_{12} = -2i\exp\{(G - i\Omega)t/2\}\left(\frac{V_{12}}{2v}\right)\sin(vt), \quad (29b)$$

$$I_{21} = -2i\exp\{-(G - i\Omega)t/2\}\left(\frac{V_{12}}{2v}\right)\sin(vt), \quad (29c)$$

$$I_{22} = \exp\{-(G - i\Omega)t/2\}\left[\cos(vt) + \left(\frac{G - i\Omega}{2v}\right)\sin(vt)\right], \quad (29d)$$

with "precession" frequency v [see (15c)]

$$v = \sqrt{\frac{(\Omega + iG)^2}{4} + V_{12}^2} \quad (30a)$$

and system decay parameter

$$G = (\gamma_1 - \gamma_2)/2. \quad (30b)$$

From relations (29a–d) and (30a) one can infer that $I_{ij}(-\Omega) = I_{ij}(\Omega)^*$, and therefore the magnitude of each element I_{ij} is symmetric about the resonance center $\Omega = 0$; the predicted resonance frequency is exactly $\omega = \omega_0$. For the case where one state is initially populated and the other unpopulated, the occupation or transition probabilities $|c_i(t)|^2$ ($i = 1, 2$) as a function of frequency produce symmetric oscillatory lineshapes. Note that only the difference in decay rates (as expressed by G), and not the individual decay constants, enter each element I_{ij}. Consequently, the rotating-field resonance line shape for transitions induced between decaying states of the same lifetime is identical to that for stable levels except for the diminution in overall intensity due to the decay matrix. This is, of course, not the case if the coupled states have different lifetimes.

Let us return to the initial problem of solving the Schrödinger equation for a two-level atom in an oscillating field and determine the effects of the antiresonant component. We start again with the exact Schrödinger equation, (25), in the interaction representation and transform it—not into the "rotating" reference frame—but into the frame of the antirotating component of the oscillating field

$$\Psi_A = \begin{pmatrix} e^{-i\tilde{\Omega}t/2} & 0 \\ 0 & e^{i\tilde{\Omega}t/2} \end{pmatrix} \Psi_I \tag{31}$$

by a transformation analogous to (27a). This leads to a Schrödinger equation in which the Hamiltonian can be decomposed into a time-independent part and a part that oscillates at 2ω. Following the procedure outlined in the previous section, one next diagonalizes the time-independent part and re-expresses the equation in a basis of its eigenstates. At this point, the substantive part of the antiresonant interaction has been captured, so to speak, in the time-independent eigenvalues and eigenvectors, and one transforms the Schrödinger equation to the reference frame of the rotating component by setting

$$\Psi_R = \begin{pmatrix} e^{i\omega t} & 0 \\ 0 & e^{-i\omega t} \end{pmatrix} (C_A^{-1} \Psi_A), \tag{32a}$$

where the matrix

$$C_A = \begin{pmatrix} 1 & -\kappa_A \\ \kappa_A & 1 \end{pmatrix} \tag{32b}$$

effects the diagonalization in the antirotating frame. The element κ_A, defined by

$$\kappa_A = \frac{\vartheta}{\sqrt{1 + \vartheta^2} + 1}, \tag{33a}$$

where

$$\vartheta = \frac{2iV_{12}}{G + i\tilde{\Omega}}, \tag{33b}$$

will ordinarily be a small quantity (since $V_{12} \ll \omega + \omega_0 \sim 2\omega_0$) and serves as a weak coupling parameter characterizing the interaction of the atom with the antiresonant part of the field. The resulting Schrödinger equation in the rotating frame takes the form

$$\left[\begin{pmatrix} \varepsilon_A^- + i\omega & \dfrac{-iV_{12}}{1 + \kappa_A^2} \\ \dfrac{-iV_{12}}{1 + \kappa_A^2} & \varepsilon_A^+ - i\omega \end{pmatrix} \right.$$

$$\left. + \begin{pmatrix} \left(\dfrac{-2iV_{12}\kappa_A}{1 + \kappa_A^2}\right)\cos(2\omega t) & \left(\dfrac{iV_{12}\kappa_A^2}{1 + \kappa_A^2}\right)e^{4i\omega t} \\ \left(\dfrac{iV_{12}\kappa_A^2}{1 + \kappa_A^2}\right)e^{-4i\omega t} & \left(\dfrac{2iV_{12}\kappa_A}{1 + \kappa_A^2}\right)\cos(2\omega t) \end{pmatrix} \right] \Psi_R = \frac{d}{dt}\Psi_R, \tag{34a}$$

in which

$$\varepsilon_A^{\pm} = -\tfrac{1}{4}(\gamma_1 + \gamma_2) \mp \tfrac{1}{2}\sqrt{(G + i\tilde{\Omega})^2 - 4V_{12}^2} \qquad (34b)$$

are the eigenvalues obtained in the antirotating frame.

It should be emphasized that the oscillating field analysis to this point is exact; no approximations have yet been made, and (34a) contains all the information of the initial Schrödinger equation (24b). The virtue of (34a) lies in the following considerations. Not only is the parameter κ_A usually small—diminishing steadily in magnitude across the resonance profile roughly as the inverse of the applied frequency—but it multiplies highly oscillatory terms in 2ω and 4ω which contribute negligibly close to resonance. Thus, to a good approximation, one can ignore the time-dependent part of (34a) and apply the theory of the preceding section to solve the residual time-independent equation. The details of obtaining the resulting closed form solution, which is somewhat complicated, are left to the original literature [18], and the solution, itself, designated by the interaction elements J_{ij} $(i, j = 1, 2)$ corresponding to the rotating-field expressions (29a–d), is given in the Appendix to this chapter.

Let us examine, however, the "precession" frequency μ,

$$\mu = \sqrt{\left[\left(\frac{\tilde{\Omega} - iG}{2}\right)(1 + \vartheta^2)^{1/2} - \omega\right]^2 + \frac{V_{12}^2}{(1 + \kappa_A^2)^2}}, \qquad (35a)$$

obtained by diagonalizing the time-independent part of (34a). This is the frequency at which transitions between the two states are induced and from which the observed central frequency of an electric resonance line shape can be determined. By expanding the inside radical in a Taylor series to order ϑ^2 and neglecting the small term κ_A^2 in the denominator of the second term, one can reduce (35a) to the form

$$\mu \sim \sqrt{\frac{[(\Omega - \delta\Omega) + i(G - \delta G)]^2}{4} + V_{12}^2} \qquad (35b)$$

directly comparable to the corresponding expression for v, (30a) of the rotating-field theory. One would then find that the oscillating field gives rise to an apparent shift in resonance frequency of

$$\delta\Omega = \frac{2V_{12}^2\tilde{\Omega}}{(\tilde{\Omega}^2 + G^2)} \qquad (36a)$$

and change in the decay-rate difference of

$$\delta G = \frac{2V_{12}^2 G}{(\tilde{\Omega}^2 + G^2)}. \qquad (36b)$$

At resonance ($\tilde{\Omega} = 2\omega_0$) the frequency shift (36a) reduces in the case of stable states to (V_{12}^2/ω_0) as derived by Bloch and Siegert [21]. (Equation (36a) differs from the result derived by Willis Lamb [22], but has been confirmed by exact numerical integration of the Schrödinger equation [23].)

An oscillatory field can do more than just shift the resonance frequency and decay-rate difference, however. Particularly in the case of transitions induced between unstable states, the entire line shape—i.e., either the occupation or transition probability as a function of frequency—can exhibit additional structure, a high-frequency modulation of the basic line shape derived by the rotating-field approximation. An example of this is illustrated in Figure 5.5 for transitions induced between an initially populated hydrogen $4S_{1/2}$ state $(c_1(0) = 1)$ and an initially unpopulated $4P_{1/2}$ state $(c_2(0) = 0)$ where the lifetimes are, respectively, $\tau_1 = 232$ ns and $\tau_2 = 12.4$ ns. In the case of the longer-lived S state, the resonance profile (not shown), calculated from the expressions

$$P_1(t) = |\langle 1|\Psi_{OF}(t)\rangle|^2 = |J_{11}(t)|^2 \exp(-\gamma_1 t), \tag{37a}$$

$$P_1^0(t) = |\langle 1|\Psi_{RF}(t)\rangle|^2 = |I_{11}(t)|^2 \exp(-\gamma_1 t), \tag{37b}$$

is virtually the same for the rotating-field (Ψ_{RF}) and oscillating-field (Ψ_{OF}) solutions. This is not the case in a corresponding comparison of the oscillating-field and rotating-field profiles of the short-lived P state

$$P_2(t) = |\langle 2|\Psi_{OF}(t)\rangle|^2 = |J_{21}(t)|^2 \exp(-\gamma_2 t), \tag{37c}$$

$$P_2^0(t) = |\langle 2|\Psi_{RF}(t)\rangle|^2 = |I_{21}(t)|^2 \exp(-\gamma_2 t), \tag{37d}$$

which differ markedly outside the immediate vicinity of resonance. In all cases the resonance line shapes determined from the oscillating-field solution and numerical integration of the Schrödinger equation are nearly indistinguishable.

The modification of the rotating-field resonance line shape by the counter-rotating component of the oscillating field can be approximated by retaining in the expression for J_{21} (equation (A2c) in the Appendix) only factors linear in κ_A. This leads to the approximate relation

$$P_2(t) \approx P_2^0(t) + \frac{1}{\overline{\Omega}} \text{Im}\left((1 - \exp(-2i\omega t)) \left(\frac{2v \cos(vt)}{\sin(vt)} - G + i\Omega \right) \right) P_2^0(t). \tag{38}$$

Thus, the total line shape is the sum of the rotating-field line shape and a smaller term inversely proportional to $\omega + \omega_0$ and modulated at the frequency 2ω (which accounts for the separation $\Delta\omega$ between modulation maxima in Figure 5.5 given by $(\Delta\omega)t = \frac{1}{2}$).

Those familar with experimental electric or magnetic-resonance line shapes may wonder why the high-frequency modulation is not usually evident. First, this additional structure is most pronounced in the case of short-lived states, whereas the observed signal ordinarily results from transitions

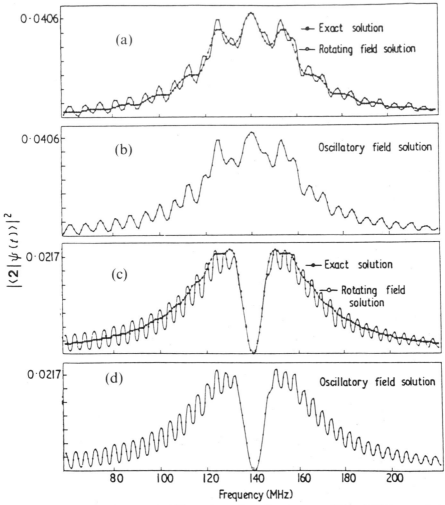

Figure 5.5. Occupation probability of a short-lived state ($|2\rangle = |4P\rangle$) coupled to a long-lived state ($|1\rangle = |4S\rangle$) as a function of driving frequency for theoretical parameters: $V_{12}/\omega_0 = \frac{5}{140}$; $c_1(0) = 1$, $c_2(0) = 0$; $\gamma_1 = 4.35 \times 10^6$ s^{-1}, $\gamma_2 = 80.6 \times 10^6$ s^{-1}; interaction time $t = 80$ ns (lineshapes (a) and (b), 125 ns (lineshapes (c) and (d). (Adapted from Silverman and Pipkin [18].)

between long-lived states, as, for example, in the pioneering experiments of Rabi and Lamb. Second, if there is a sufficiently wide dispersion in the atomic velocities, as in the case of a thermal beam, the transition probability must be averaged over the interaction time and this tends to smooth out the resonance line shape. It is worth noting, however, that the modulation is largely unaffected by the phase of the field. In the preceding analysis the initial phase of the oscillating field was arbitrarily chosen to be zero. For an experimental configuration such as an atomic beam, however, where atoms are continually passing through the interaction region, the initial phase of the field sampled by different atoms spans the full range of possibilities from 0 to 2π. To correspond to the experimental signal, therefore, theoretical expressions for occupation or transition probabilities must be averaged over the phase of the field. The phase, as will be shown in Section 5.5, appears in a phase factor multiplying the off-diagonal elements of the matrix (I_{ij}) or (J_{ij}). The probability of transition out of an initially pure state is totally insensitive to the phase of the field. This is not to say, however, that phase plays no rôle. In the following sections important experimental consequences of a sharp relative phase will be explored further in experimental configurations giving rise to quantum interference.

A detailed examination of the variation of the oscillating-field solution with diverse experimental parameters (frequency and power of the applied field and duration of interaction) must again be left to the original literature. However, it is worth mentioning one further unusual, or perhaps unanticipated, fact concerning the response of the atom to an increasing applied field strength (as represented by the matrix element V_{12}) in the case of decaying states.

First note that in the absence of spontaneous decay the occupation probabilities at resonance predicted by the rotating-field approximation reduce to the familiar results

$$P_1(t) = \cos^2(V_{12}t), \tag{39a}$$

$$P_2(t) = \sin^2(V_{12}t), \tag{39b}$$

which oscillate between 0 and 1 with a recurrence relation $(\Delta V_{12})t = \pi$. From these expressions it is obvious that, for a fixed interaction time, the maximum probability for transition *out* of one state occurs at the field strength at which the transition *into* the other state is also maximum. This must be so, for there are but two states, and it is obvious from relations (39a, b) that total probability is conserved.

Figure 5.6 shows the variation in probabilities $P_1(t)$ and $P_2(t)$ obtained from the rotating-field approximation for the resonant coupling ($\omega = \omega_0$) of hydrogen $4S_{1/2}$ and $4P_{1/2}$ states whose lifetimes were given earlier; the duration of the interaction with the field is 50 ns. Since the exact and oscillating-field calculations yield virtually identical results, they are not shown. When $V_{12} = 0$, and therefore no transitions are induced, the probabilities are seen to be $P_1 = 0.8$ and $P_2 = 0$ as determined by the initial conditions. With increasing strength of the applied field, the probability of remaining in the longer-lived state 1 rapidly decreases and then oscillates between zero and a constant maximum value less than unity with the same recurrence relation $(\Delta V_{12})t = \pi$. The occupation probability of the initially unpopulated state 2 reaches a maximum at a certain value of V_{12}, then

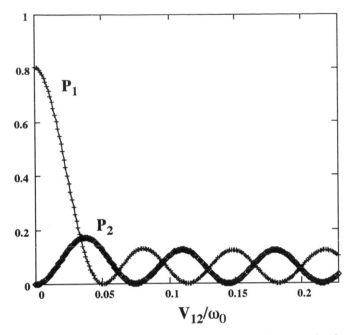

Figure 5.6. Variation of occupation probabilities P_1, P_2 with strength of atom-field coupling for a long-lived $4S$ state and short-lived $4P$ state, respectively. The interaction time is 50 ns; initial amplitudes and decay rates are the same as for Figure 5.5.

exhibits analogous oscillatory behavior although with a periodicity that depends on the decay parameter G.

As clearly illustrated in Figure 5.6, however, for transitions between unstable states a maximum quenching of one state need no longer occur at the same field strength as a maximum pumping of the other. This can be readily inferred from the theoretical resonance expressions analogous to (39a, b)

$$P_1(t) = \exp\left(-(\gamma_1 + \gamma_2)\frac{t}{2}\right)(\cos Xt - (G/2X)\sin Xt)^2, \qquad (40a)$$

$$P_2(t) = \exp\left(-(\gamma_1 + \gamma_2)\frac{t}{2}\right)\frac{V_{12}^2}{X^2}\sin^2 Xt, \qquad (40b)$$

where

$$X = \sqrt{V_{12}^2 - (G/2)^2}. \qquad (40c)$$

The reason for this is that an unstable two-level atom is not really a two-level system, being coupled by the vacuum electromagnetic field to all states of lower energy and appropriate symmetry. Indeed, as readily verified from Figure 5.6 and (40a, b) or (23b), the sum of the probabilities at resonance—or indeed at any frequency—is not unity.

5.4. Resonance and Coherent States: The Tell-Tale Mark of a Quantum Jump

All physicists at some time during their study of the quantum theory of angular momentum undoubtedly encounter the seemingly peculiar property of spinors that a rotation of 4π, rather than 2π, radians is required to return them to their original state. A rotation of 2π radians, which intuitively ought to correspond to no rotation at all, multiplies the spinor wave function by -1, or equivalently by the phase factor $e^{i\pi}$. Theoretically, this follows from the form of the unitary operator $U(\theta) = \exp(-i\mathbf{J} \cdot \mathbf{n}\theta)$ which rotates a state vector of angular momentum $\mathbf{J}$ (in units of $\hbar$) by an angle θ about the direction specified by unit vector $\mathbf{n}$. If the magnitude of $\mathbf{J}$ (or, more precisely, the angular momentum quantum number) is an odd half integer (e.g., $J = \frac{1}{2}, \frac{3}{2}$, etc.), then $U(2\pi) = -1$ independent of the rotation axis.

In so far as one is discussing the properties of an abstract mathematical object, the above rotational property is not at all disturbing. Mathematics need satsfy no criteria imposed by the real world for its justification. But physics clearly must, and if spinors are to be suitable representations of real fermionic systems, then it becomes a legitimate question to ask whether the consequences of rotating a spinorial system by 2π are observable.

The pedagogical literature has not been encouraging in this regard. Dirac, himself, asserted as a "general result" in his classic treatise on the principles of quantum mechanics that: "... the application of one revolution about any axis leaves a ket unchanged or changes its sign. A state, of course, is always unaffected by the revolution, since a state is unaffected by a change of sign of the ket corresponding to it" [24]. This sentiment has been often repeated in physics textbooks. One finds, for example, in one well-known book [25]: "That the rotation of a ket through 2π does not give the same ket raises no difficulty of principle so long as no observable effect is produced." Underlying all such remarks is the basic idea that the result of any measurement is representable by expectation values bilinear in the wave function. Therefore two wave functions differing only in overall sign cannot lead to different physical predictions.

There is, in fact, nothing incorrect in the above assertions. Indeed, the rotation of an isolated system—measuring apparatus included—does not lead to experimental consequences. This is not, however, what one ordinarily means by a rotation. Only a part of a system can be rotated; part must remain fixed to provide a reference against which the rotation is to be measured. The global phase change resulting from the former process has no experimental counterpart and therefore no physical implications. The latter process, the rotation of a spinor-characterized portion by 2π radians with respect to a fixed portion of an encompassing larger system, *does* have physical implications, as demonstrated in particular by a number of clear experiments [26].

In the first, and perhaps best known, of these experiments the rotation of neutrons, which are spin-$\frac{1}{2}$ particles, was observed by means of neutron interferometry [27, 28]. In basic outline, a beam of unpolarized neutrons was coherently divided at a single-crystal beam splitter, one component passing between the poles of a magnet, the other component passing directly through field-free space to the analyzer where both components were subsequently recombined. As a result of the two-beam interference, the neutron intensity transmitted by the analyzer exhibited oscillations as a function of the magnetic field strength.

Considered classically, each neutron is a small magnetic dipole and therefore undergoes Larmor precession in a magnetic field. For appropriate values of magnetic field and interaction time, any desired angle of precession can be realized. Quantum mechanically, the precession angle corresponds to the relative phase shift between the two components of the total wave function. Thus, from the periodicity of the observed oscillations in neutron intensity, such as shown in Figure 5.7, one could conclude

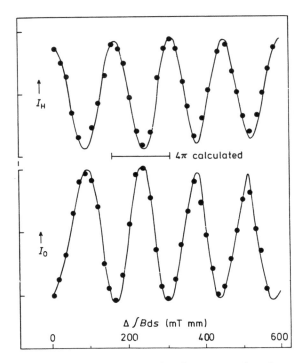

Figure 5.7. Demonstration of spinor rotation by neutron interferometry. A single-crystal interferometer splits an incident neutron beam into two components, one of which passes between the poles of a magnet. The two beams recombine at the rear face of the interferometer and emerge in a forward beam (O) and a deviated diffracted beam (H). The intensity of each, determined by neutron counting, manifests oscillations with a 4π periodicity as a function of magnetic field. (Adapted from Rauch et al. [27].)

that a neutron returned to its original state after precessing through 4π (rather than 2π) radians.

Although seemingly straightforward, the neutron rotation experiments give rise to some conceptual difficulties of interpretation [29]. At the root of the difficulty was the Heisenberg uncertainty principle; for fermions the simultaneous observation of the relative rotation of the spins in the two beams and of the interference pattern is incompatible. In the context of an interference experiment one can never know whether a particular neutron followed a classical path through the magnetic field or the field-free region. Although one can measure the rotation of neutrons along one path relative to the rotation of neutrons along the other, an intrusive measure ment of this kind would destroy the interference pattern. Thus, according to Byrne, who first called attention to this complication, "... the notion of relative rotation ceases to have a meaning as it corresponds to nothing which is measurable."

The objection derives from examination of the density matrix of the beam of neutrons. For unpolarized neutrons the experimental possibilities are exhausted by observation of the interference pattern and determination of the fringe visibility. If the neutrons are polarized, one can determine the degree of polarization parallel to and transverse to the magnetic field. The former, the longitudinal polarization, is unchanged by division and subsequent recombination of the wave function. From Byrne's analysis, it is the rotation of the transverse polarization that is measured by neutron interferometry, and this observable corresponds to either a rotation or nutation of the particles in the *recombined* beam relative to those in the *incident* beam. In no case, however, does one observe the relative rotation of particles in the *split* beams.

The reinterpretation of the neutron interferometry experiments may at first glance seem like quibbling over semantics. It is not, however, since a similar situation does not arise in the case of massive bosons which always have a transverse spin component ($m_S = 0$) whose complex amplitude is unaffected by passage through the magnetic field. The transverse polarization in this case is expressible as the incoherent sum of a fixed polarization and a rotated polarization. Experimentally, the observed angle of rotation corresponds to the angle between *two* orientations of the analyzing axis of a filter giving maximum transmission of the particles—in contrast to fermions for which there is but *one* transmission axis. Thus, the concept of relative spin rotation retains a meaning for massive bosons in the context of an interference experiment. (The rotation of a boson, however, is not of particular conceptual interest, since it returns to its original state with the usual angular periodicity of 2π.)

In the preceding chapter we have examined various facets of the physics of laser-induced quantum beats. It will be recalled that, from the perspective of spectroscopy, one of the principal advantages of this technique was to eliminate the need for external radiofrequency or microwave fields.

Spectroscopy aside, however, the application of an oscillating field to states prepared in a coherent linear superposition permits one to determine the relative phase of different components of the wave function—a result of interest in its own right. One noteworthy application pertinent to the above discussion on spinors is that under appropriate circumstances a quantum-beat resonance experiment can reveal the 4π periodicity of spinor rotation. Actually, the experiment is of more general scope, for, it does not require two beams of spin-$\frac{1}{2}$ particles. Furthermore, since particles are neither physically rotated nor spatially diffracted, no conceptual difficulties arise concerning the classical interpretation of rotation in a quantum mechanical context. Indeed, examined from a purely quantum perspective, what the experiment directly shows is the perhaps more surprising fact that an atom which has undergone a transition from one state to a second and then back again to the first may still be experimentally distinguishable from an identical atom which has undergone no transition at all [30].

Let us consider a beam of atoms with ground state $|0\rangle$ and three close-lying excited states $|1\rangle$, $|2\rangle$, $|3\rangle$ labeled in order of increasing energy (expressed in units of $\hbar$) $\omega_1 < \omega_2 < \omega_3$ such that the energy (i.e., frequency) intervals ω_{32} and ω_{21} (where $\omega_{ij} = \omega_i - \omega_j$) are unequal. The configuration of states is shown in Figure 5.8. At a fixed time, $t = 0$, a laser pulse of mean frequency ν_0 where $\omega_1 < \nu_0 < \omega_2$ and bandwidth $\Delta\nu$ satisfying $\omega_{21} < \Delta\nu < \omega_{32}$ prepares the beam in a linear superposition of states $|1\rangle$ and $|2\rangle$ with respective amplitudes a_1 and a_2. Were the atoms to evolve freely in time under the Hamiltonian H_0 whose eigenvalues are the frequencies

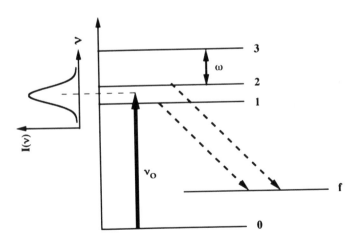

Figure 5.8. Diagram of a quantum beat resonance experiment illustrating the observable effect of a cyclic quantum transition. A laser pulse of intensity $I(\nu)$ centered on frequency ν_0 excites an atom into a linear superposition of states 1 and 2. A rf field of frequency ω induces the cyclic transition $2 \to 3 \to 2$. The associated phase change is observable in the quantum beats of the radiative decay to lower level f.

ω_i, the state vector at some subsequent time t would take the form

$$\Psi(t) = e^{-iH_0 t}\Psi(0) = a_1 e^{-i\omega_1 t}|1\rangle + a_2 e^{-i\omega_2 t}|2\rangle. \tag{41}$$

In what follows the ground state plays no role and has been eliminated from the above superposition. Also, each amplitude should in principle be multiplied by an exponential damping factor of the form $\exp(-t/2\tau_i)$ where τ_i is the mean lifetime of the ith state. However, in the case of a sufficiently long, although finite, lifetime for whith $1/\tau \ll \omega_{ij}$, so many oscillations between superposed states occur within one e-folding time, that it is permissible to simplify the ensuing mathematics without sacrificing essential physical ideas by omitting the decay factors.

Following the analysis of Chapter 4, one can express the radiation intensity of polarization ε emitted at time t as a result of a spontaneous transition to a lower state $|f\rangle$ by

$$I(t) = |\langle f|\mu_E \cdot \varepsilon|\Psi(t)\rangle|^2, \tag{42}$$

where μ_E is the electric dipole operator. Thus, substitution of (41) into (42) leads to an optical signal with harmonic component, the quantum beat, at frequency ω_{21}. For the production of the beat it is essential that the transition matrix elements $D_{ij} \equiv \langle i|\mu_E \cdot \varepsilon|j\rangle$ of both initial states ($i = 1, 2$) to the same final state ($j = f$) be nonzero.

The atoms in the beam, however, are *not* permitted to evolve freely, but are subjected instead to an oscillating electric field of adjustable frequency ω and arbitrary phase φ giving rise to the perturbation $V(t)$ (in units of angular frequency). The oscillation frequency is chosen so as to be in the vicinity of ω_{32} and to differ appreciably from ω_{21}—i.e., to couple states $|3\rangle$ and $|2\rangle$ without appreciably coupling states $|2\rangle$ and $|1\rangle$. Alternatively, the interaction $V(t)$ may be of such symmetry that only the element $V_{32} = \langle 3|V_0|2\rangle$ is non-null (where V_0 is the amplitude of the oscillatory potential $V(t)$). Thus, in regard to this interaction, the atom behaves like a two-state system, since transitions can be induced only between states $|3\rangle$ and $|2\rangle$. The state of the atom at time t, whose initial state is given by (41) with $t = 0$, is obtained from solution of the time-dependent Schrödinger equation with potential $V(t)$. The equation immediately reduces to two uncoupled equations: one for a_1, whose time dependence is already given in (41), and a second for the coupled amplitudes a_2 and a_3.

For a weak perturbation—i.e., where the matrix elements of V are small compared to state eigenfrequencies—one can apply the rotating-field approximation developed in the preceding section. Then, with neglect of the small shift in resonance frequency and radiative decay rates during passage through the field, the solution to the two-state equation can be written in the form of a spinor rotation followed by free evolution

$$\begin{pmatrix} b_2 \\ b_3 \end{pmatrix} = \exp(-iH_0 t)\exp\left(\frac{i\sigma_3\Omega t}{2}\right)\exp\left(\frac{-i\mathbf{n}\cdot\boldsymbol{\sigma}\theta}{2}\right)\begin{pmatrix} a_2 \\ 0 \end{pmatrix}. \tag{43}$$

Here, $\boldsymbol{\sigma} = (\sigma_1, \sigma_2, \sigma_3)$ is again the vector of Pauli spin matrices,

$$\Omega = (\omega - \omega_{32}) \tag{44a}$$

is the deviation from resonance, and

$$\theta = 2\left[|V_{32}|^2 + \frac{\Omega^2}{4}\right]^{1/2} t \equiv 2\Omega_r t \tag{44b}$$

is the "rotation" or "precession" angle (with corresponding angular frequency Ω_r). The unit vector $\mathbf{n}$ specifying the rotation axis has components (n_1, n_2, n_3) given by

$$\mathbf{n} = \left[\frac{V_{32} \cos \varphi}{\Omega_r}, \frac{V_{32} \sin \varphi}{\Omega_r}, \frac{\Omega}{2\Omega_r}\right]. \tag{45}$$

For simplicity V_{32} has been taken to be real-valued. Note, too, that the arbitrary phase φ of the rf field is contained only in the components n_1 and n_2, and this feature is unchanged should V_{32} be complex-valued.

At resonance, $\Omega = 0$, and the rotational parameters reduce to the simple expressions

$$\mathbf{n} = (\cos \varphi, \sin \varphi, 0) \, \mathrm{sgn}(V_{32}), \tag{46a}$$

$$\theta = 2|V_{32}|t. \tag{46b}$$

The signum function $\mathrm{sgn}(x)$ returns the sign of its argument. By adjustment of the oscillating-field strength (to which V_{32} is proportional) and duration t of the atom-field interaction, the angle θ can be appropriately selected. Equations (46a, b) show that under resonant conditions a variation in the field strength does not alter the rotation axis, since $\mathbf{n}$ is independent of $|V_{32}|$. Moreover, for rotations of an integer multiple of 2π radians, the state of the system is independent of $\mathbf{n}$ and therefore of the arbitrary phase φ.

From (43) through (46)—with application of the identity in relation (7a)—it follows that at resonance a rotation of $\theta = \pi$ results in all atoms initially in state $|2\rangle$ being driven into state $|3\rangle$

$$\left(\begin{matrix} b_2 \\ b_3 \end{matrix}\right)^{\Omega=0}_{\theta=\pi} = \exp(-iH_0t)\left(\begin{matrix} 0 \\ -ia_2e^{i\varphi} \end{matrix}\right), \tag{47a}$$

whereas a rotation of $\theta = 2\pi$ corresponds to a transition from $|2\rangle$ to $|3\rangle$ and back again to $|2\rangle$

$$\left(\begin{matrix} b_2 \\ b_3 \end{matrix}\right)^{\Omega=0}_{\theta=2\pi} = -\exp(-iH_0t)\left(\begin{matrix} a_2 \\ 0 \end{matrix}\right). \tag{47b}$$

The occurrence of the minus sign is normally not observed since it is not revealed by any bilinear products of the wave function. In the present experiment, however, this sign change is important and will be seen to have experimental consequences. A minimal resonant rotation of $\theta = 4\pi$, which corresponds to two cyclic transitions between $|2\rangle$ to $|3\rangle$, is required to return the system to the same state as if no transition had occurred.

We now return to a description of the full three-state system. At the time of emergence from the oscillating field, the state vector of the atom is generally given by

$$\Psi(t) = \exp(-iH_0 t)\left[a_1|1\rangle + \exp(i\sigma_3\Omega t/2)\exp(-i\mathbf{n}\cdot\boldsymbol{\sigma}\theta/2)\binom{a_2}{0}\right], \quad (48a)$$

which, when expanded, takes the explicit form

$$\Psi(t) = e^{-iH_0 t}[a_1|1\rangle + a_2 e^{i\Omega t/2}(\cos(\theta/2) - in_3\sin(\theta/2))|2\rangle$$
$$- ia_2 e^{-i\Omega t/2}(n_1 + in_2)\sin(\theta/2)|3\rangle]. \quad (48b)$$

The operator $\exp(-iH_0 t)$ acts on each basis ket $|i\rangle$ to generate the phase factor $\exp(-i\omega_i t)$ which occurs under free evolution. Thus, after passage through the oscillating field, free evolution for a time T is accounted for by replacing t by $t + T$ in the first factor of (48b). One can also readily verify that the total state vector is still properly normalized to $|\Psi|^2 = 1$ as expected (spontaneous decay having been neglected).

At an arbitrary time T after emergence from a *resonant* field the atom is characterized by the state vector

$$\Psi_{\Omega=0}(t + T) = \exp(-iH_0(t + T))$$
$$\times [a_1|1\rangle + a_2\cos(\theta/2)|2\rangle - ia_2 e^{i\varphi}\sin(\theta/2)|3\rangle]. \quad (49)$$

At this point the spontaneous emission from the atom plays a role, for it furnishes the optical signal to be observed. The general expression for the resonant atomic fluorescence of polarization ε emitted at time $t + T$, obtained by substituting (49) into (42), is somewhat complicated (with quantum beats at frequencies ω_{21}, ω_{31}, and ω_{32}) and need not be given explicitly. Of particular significance, however, is the result that for "rotations" of 0 and 2π, the quantum-beat signals at frequency ω_{21} are independent of the arbitrary initial phase of the oscillating field and differ in relative phase by 180° as shown below

$$I_{\Omega=0}^{0,2\pi}(t + T) = |a_1|^2|D_{f1}|^2 + |a_2|^2|D_{f2}|^2$$
$$\pm 2\,\mathrm{Re}\{a_1 a_2^* D_{f1}D_{f2}^*\exp(i\omega_{21}(t + T))\}, \quad (50)$$

where the upper and lower signs correspond to 0, 2π, respectively.

Experimentally, one can measure the above fluorescent intensities in the traditional way as a function of T for fixed amplitude of the oscillating field. The effect of the sign change can be enhanced by taking the difference signal

$$S(t + T) = I_{\Omega=0}^0(t + T) - I_{\Omega=0}^{2\pi}(t + T)$$
$$= 4|a_1 a_2^*||D_{f1}D_{f2}^*|\cos(\omega_{12}t + \beta) \quad (51)$$

to obtain ideally a harmonic transient with 100% contrast. (The signal, of course, is exponentially damped when state decay is explicitly included in the analysis.) The phase β depends on the interaction time t and on constant phase factors arising from possibly complex-valued matrix elements and initial amplitudes. Thus, the distinction between a cyclic transition between two states ($\theta = 2\pi$) and no transition at all is revealed by a tell-tale minus sign in the wave function.

To show explicitly the 4π periodicity of the transition one can modify the foregoing procedure to measure, at fixed time $t + T$, the fluorescence as a function of the oscillating-field strength or equivalently θ. Since the atoms passing through the field experience all values of the initial phase on which the signal in general depends, one must in this case average $I(t + T)$ over the range $0 \leq \varphi \leq 2\pi$. Assuming that the atomic states have a well-defined parity—which is valid upon neglect of small effects on the electron wave function due to weak nuclear interactions—the states $|2\rangle$ and $|3\rangle$ must have opposite parities to be coupled by an oscillating electric field. Thus, for an electric dipole decay transition to the lower state $|f\rangle$, one of the matrix elements D_{f2} or D_{f3} must be null. With $|f\rangle$ chosen so that $D_{f3} = 0$, for example, the calculated phase-averaged intensity at resonance can be shown to be

$$\langle I_{\Omega=0}(t + T)\rangle_\varphi = |a_1|^2 |D_{f1}|^2 + |a_2|^2 |D_{f2}|^2 \cos^2(\theta/2)$$
$$+ 2 \operatorname{Re}\{a_1 a_2^* D_{f1} D_{f2}^* \exp(i\omega_{21}(t + T))\} \cos(\theta/2). \quad (52a)$$

The signal represented by (52a) is then processed by taking pointwise over a range of θ the difference

$$S(\theta) = \langle I_{\Omega=0}(\theta)\rangle_\varphi - \langle I_{\Omega=0}(\theta + \pi)\rangle_\varphi$$
$$= 4 \operatorname{Re}\{a_1 a_2^* D_{f1} D_{f2}^* \exp(i\omega_{21}(t + T))\} \cos(\theta/2) \quad (52b)$$

which results in a signal explicitly showing the 4π periodicity of spinor rotation.

It is worth noting expressly that the phase change under a cylic transition is observable in this experiment precisely because it is *not* a global phase, but rather a relative phase between the components of the wave function characterizing states $|2\rangle$ and $|1\rangle$. The latter is unaffected by the external oscillating field and thereby assumes the role of a reference state in much the same way in which phase information of light scattered from an object is recorded holographically.

As a final consideration on spinors, I would like to point out an amusing, if not intellectually mystifying, demonstration that I have often used before lecture audiences to illustrate the spinor rotational property concretely by means of a classical object—rather than a purely quantum system like an electron or neutron. Invented originally by Dirac, as far as I am aware, and referred to as a "spinor spanner," the apparatus consists of a spanner (or wrench) fastened by three parallel cords to two vertical walls as shown in Figure 5.9. If the spanner is rotated 360° about

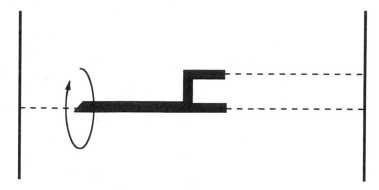

Figure 5.9. The "spinor spanner." Rotating the spanner around its axis by 360° entangles the cords which can be made to untangle by a further 360° rotation in the same sense. The initial impression following a 720° rotation, however, is that the cords are even more entangled than before.

its axis, the cords become twisted, and no maneuvering of the object or cords—short of a counter-rotation or cutting of the cords—can undo the tangle. If the spanner is now turned another 360° in the same sense, the entanglement appears at first even worse than before. But the astonishing thing is that by looping the cords around the spanner in an appropriate manner—the orientation of the spanner all the while kept fixed—one can untangle the snarl and recover the original configuration of the system as shown in the figure. Why a 4π rotation is equivalent to the identity operation in this case is an instructive exercise in topology [31] which cannot be pursued further in this book. The demonstration itself, however, is easy enough to construct and will undoubtedly fascinate the reader as it has for many years the author.

5.5. Quantum Interference in Separated Oscillating Fields

In the discussion of atomic state selection in Section 5.1, the various oscillators that furnished the distinct spectroscopy and quenching fields were uncorrelated, i.e., they had in general different frequencies and no well-defined relative phase. A configuration of two (or more) sequential oscillating fields, such as illustrated in Figure 5.10, which *together* comprise a *single* spectroscopy region was proposed in 1949 by Norman Ramsey [32] and contributed to his award of the Nobel Prize for physics for 1989. [33] In this configuration two spatially separated interaction regions created from the *same* oscillator have identical frequencies and a well-defined relative phase which is adjustable by a phase-shifter. A sensitive spectroscopic tool for resolving resonance line shapes, this field configuration is of conceptual

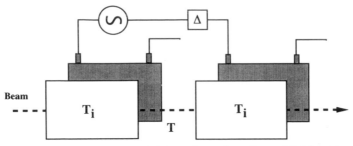

Figure 5.10. Separated oscillating field configuration. Two rf fields, driven by the same oscillator, have the same frequency and well-defined relative phase Δ. The atoms in the beam interact with each field for a time T_i with a time of transit T between fields.

interest here because it leads to a novel form of particle self-interference in the time domain.

Under the usual circumstances where an atom in a beam is not observed until after having passed through both fields, an observer can *not* know in which of the two interaction regions a transition between atomic states may have occurred. Thus, the entire configuration is again analogous to the electron two-slit experiment, except that in the present case the "slits" are not laterally spaced apertures, but longitudinally separated "windows" in time. All the same, the transition probability contains an interference term with the experimental consequence that the resonance curve exhibits an oscillatory profile narrower by nearly a factor of two than the width of the resonance produced by a single oscillatory field of the same overall length. One may have anticipated this by analogy with diffraction and interference in physical optics. The central diffraction peak of a single slit of given width is broader than the fringes of an interference pattern produced by two slits of equivalent total width. It is this interference narrowing that in part gives the separated oscillatory-field configuration its spectroscopic utility, for it is difficult to determine a precise resonance frequency from a broad line shape.

There is a second interesting feature associated with the coherently oscillating-field configuration when the coupled atomic states are unstable. As illustrated later in the section (in Figure 5.12), the greater the separation between the two interaction regions, the more oscillatory is the resulting line shape—and hence the more sharply defined is the resonant transition frequency. This is a practical consequence of the uncertainty principle. Recall that the lifetime of a quantum state is a *statistical* measure of the duration of that state. In an ensemble of similarly prepared unstable systems, some decay sooner and others last longer than the mean lifetime. Of course, the probability of surviving much longer than the mean lifetime τ diminishes exponentially as $\exp(-t/\tau)$. Thus, the probability that an atom remains in an unstable state for a period of five lifetimes before decaying is

$\exp(-5) \sim 0.0067$, or approximately seven atoms out of every thousand. By contrast, the number surviving for ten lifetimes is $\exp(-10) \sim 0.000045$ or approximately five out of one hundred thousand. Nevertheless, with a sufficiently intense initial beam, one can have a measurable population of atoms in an unstable excited state pass from the first to the second transition region if the separation between regions is not too great. The mean lifetime of this population, however, is now longer than that of the original unselected population of atoms in the same state. According to the version of the uncertainty principle concerning time intervals and energy, $(\Delta E)(\Delta t) \geq \hbar$, the longer the time interval $\Delta t = \tau$ over which a state can last, the smaller is the uncertainty ΔE, and hence the narrower will be the corresponding resonance line shape.

Let us examine quantitatively the quantum mechanics of a two-level atom with energy interval $\omega_0 = \omega_1 - \omega_2$ traversing the configuration of two separated oscillating fields. As it passes the two interaction regions, the atom experiences the electric field $E_0 \cos(\omega t + \delta_1)$ for a time T_i, field-free space for a time T, and a second electric field $E_0 \cos(\omega t + \delta_2)$ again for a time T_i. (The more general case of different field amplitudes, frequencies, and interaction times is treated in [8].) We take the two fields to be linearly polarized parallel to one another. At the time $t = 0$ of entry into the first field, the state of the atom is specified by the wave function $\Psi(0)$ which is representable as a vector, as in (2a, b), where the upper component is the amplitude of state 1 and the lower component the amplitude of state 2. Upon leaving the second field a time interval $t = 2T_i + T$ later, the state of the atom is represented by a wave function $\Psi(t)$ which can be cast in the form

$$\Psi(t) = e^{-iXt}M\Psi(0), \tag{53}$$

where the X is the operator $H_0 - i\Gamma$ in (24a) responsible for free evolution and spontaneous decay, and M is the transition matrix whose elements are sought. To determine M we will utilize the results of the preceding section and calculate the state of the atom at the entry and exit of each of the interaction regions.

At the time of emergence T_i from the first field, the wave function has evolved according to

$$\Psi(T_i) = e^{-iXT_i}K(T_i, \delta_1)\Psi(0), \tag{54}$$

where the matrix representation of X is diagonal with elements

$$X_j = \omega_j - i\gamma_j/2 \quad (j = 1, 2), \tag{55}$$

and the elements of $K(T_i, \delta_1)$, obtained in the rotating-field approximation from (29a–d), take the form

$$K_{11} = I_{11}; \quad K_{22} = I_{22}; \quad K_{12} = I_{12}e^{-i\delta_1}; \quad K_{21} = I_{21}e^{i\delta_1}. \tag{56}$$

The phase δ_1 is a distributed variable, differing from one atom to the next that enters the first field.

The time-evolution of the wave function in field-free space is generated simply by the matrix operator e^{-iXT}. To account for passage through the second interaction region one merely repeats the application of (54), replacing the initial wave function $\Psi(0)$ by the expression

$$\Psi(T + T_i) = e^{-iXT}\Psi(T_i) \tag{57}$$

to obtain

$$\Psi(t) = e^{-iXT_i}K(T_i; \varphi_2)e^{-iX(T+T_i)}K(T_i, \delta_1)\Psi(0). \tag{58}$$

The elements of the matrix $K(T_i, \varphi_2)$ have the same form as those of (56) but with the phase δ_1 replaced by

$$\varphi_2 = \omega(T + T_i) + \delta_2 \tag{59}$$

which is also a distributed variable. (In general, the phase φ in relations (56) for an atom entering the field $E_0 \cos(\omega t + \delta)$ at time t_0 is $\varphi = \omega t_0 + \delta$.)

Upon inverting the order of the inside exponential and transition matrix $K(T_i, \varphi_2)$ in expression (58), and taking account of their noncommutativity, one obtains (53) where the elements of M are

$$M_{11} = I_{11}^2 + I_{12}I_{21} \exp(G(T + T_i) - i\Theta), \tag{60a}$$

$$M_{12} = e^{-i\delta_1}[I_{11}I_{12} + I_{12}I_{22} \exp(G(T + T_i) - i\Theta)], \tag{60b}$$

$$M_{21} = e^{i\delta_1}[I_{22}I_{21} + I_{21}I_{11} \exp(-G(T + T_i) - i\Theta)], \tag{60c}$$

$$M_{22} = I_{22}^2 + I_{21}I_{12} \exp(-G(T + T_i) - i\Theta), \tag{60d}$$

and the phase Θ is

$$\Theta = \Omega(T + T_i) + (\delta_2 - \delta_1). \tag{61}$$

G is the decay function given in (30b), and Ω is the displacement from resonance, (26a). Although δ_1 and δ_2 are distributed quantities, the phase difference

$$\Delta \equiv (\delta_2 - \delta_1) \tag{62}$$

is an experimentally adjustable, well-defined parameter in the case of two interaction regions driven by the same oscillator.

The density matrix of the two-level atom is now determined by averaging the product $\Psi(t)\Psi(t)^\dagger$ over the variable δ_1 while keeping Δ constant

$$\rho(t) = \langle\Psi(t)\Psi(t)^\dagger\rangle_{\delta_1} = \langle e^{-iXt}M\rho(0)M^\dagger e^{iX^\dagger t}\rangle_{\delta_1}. \tag{63a}$$

Assuming for the sake of illustration—and because it is often the case experimentally—that the initial density matrix $\rho(0)$ represents an atom initially prepared in one of the two coupled states, for example the longer-lived state $|1\rangle$,

$$\rho(0) = \begin{pmatrix} 1 & 0 \\ 0 & 0 \end{pmatrix} \tag{63b}$$

leads to the following occupation probabilities after passage of the atom through both interaction regions:

$$\rho_{11}(t) = e^{-\gamma_1 t}|I_{11}I_{11} + I_{12}I_{21}e^{(G-i\Omega)(T+T_i)}e^{-i\Delta}|^2, \tag{64a}$$

$$\rho_{22}(t) = e^{-\gamma_2 t}|I_{22}I_{21} + I_{21}I_{11}e^{-(G-i\Omega)(T+T_i)}e^{i\Delta}|^2. \tag{64b}$$

The order of the interaction matrix elements I_{ij} in each term of (64a) and (64b) corresponds from right to left to the first and second interaction regions, respectively. Thus, in $\rho_{11}(t)$, for example, one sees that the first term characterizes an atom in state $|1\rangle$ that remains in state $|1\rangle$ after passage through both fields, whereas the second term characterizes an atom that undergoes a transition from $|1\rangle$ to $|2\rangle$ in the first field and back again from $|2\rangle$ to $|1\rangle$ in the second field. Under the conditions of the experiment there is no way to distinguish which process produced an atom emerging in state $|1\rangle$ from the second field; the total amplitude is therefore the sum of the amplitudes for these two processes, and the occupation probability $\rho_{11}(t)$ (and likewise for $\rho_{22}(t)$) contains an interference term. In keeping with the earlier heuristic argument based on the uncertainty principle, note that the relative phase between the two terms depends on the field-free interaction time T; the larger T, the more oscillatory is the exponential phase factor. Under the conditions of resonance ($\Omega = 0$) between two states of the same lifetime ($G = 0$), however, the relative phase between the two terms is independent of T; the separation between the interaction regions then has no effect except to diminish the overall intensity if the states are unstable.

Let us examine the case of exact resonance between two states of the same lifetime more closely to see better the physical effect of the relative phase Δ. When the matrix element V_{12} is greater than $G/2$, it follows from (29a–d) and (30a) that the products of interaction matrix elements I_{ij} in (64a, b) are real, and therefore the interference terms in (64a, b) are proportional to $\cos \Delta$. When $G = 0$, the precession frequency v is simply equal to the dipole matrix element V_{12}, and the occupation probability of state $|1\rangle$, for example, becomes

$$\rho_{11}(t) = e^{-\gamma_1 t}[\cos^4(V_{12}T_i) + \sin^4(V_{12}T_i) \\ - 2\cos^2(V_{12}T_i)\sin^2(V_{12}T_i)\cos\Delta]. \tag{65a}$$

For a choice of field strength and interaction time such that $V_{12}T_i = \pi/4$ radians, the above expression reduces to

$$\rho_{11}(t) = e^{-\gamma_1 t}\frac{(1 - \cos\Delta)}{2} \tag{65b}$$

and the interference term in ρ_{11} shows oscillations with 100% contrast. If the two coherently oscillating fields are in phase ($\Delta = 0$), then the atom, initially in state $|1\rangle$, emerges in state $|2\rangle$ with 100% probability. On the other hand, if the two fields are 180° out of phase ($\Delta = \pi$), the emerging atom will be found in the original state $|1\rangle$ with 100% probability.

Under conditions where the frequency is not exactly at resonance or the states are not stable, the interference term still diminishes the $|1\rangle$ component of the wave function for in-phase oscillating fields and augments this component for fields oscillating out-of-phase. However, the contrast in these cases may be too low for different choices of Δ to be directly noticeable in the variation of ρ_{11} or ρ_{22} with oscillation frequency. One can then isolate the interference term, as was done in the preceding section for a quantum-beat signal, by measuring the difference in occupation probability

$$S_i(\Delta_1, \Delta_2) = \rho_{ii}(\Delta_1) - \rho_{ii}(\Delta_2) \tag{66a}$$

for two appropriately chosen values of the relative phase Δ. One loses, of

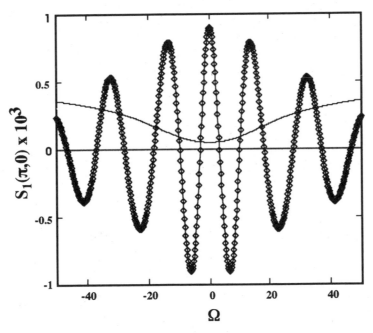

Figure 5.11. Isolation of quantum interference effect produced by separated coherently oscillating fields. The oscillatory curve is the difference in occupation probabilities of a long-lived state (3S coupled to 3P) for relative phase $\Delta = \pi, 0$, as a function of the frequency interval $\Omega = \omega - \omega_0$. The interference is not visible in the frequency variation of the individual occupation probabilities $\rho_{11}(\Delta = \pi, 0)$ (superposed curve). Theoretical parameters are: $V_{12} = 2\pi \times 6\,\text{MHz}$; $c_1(0) = 1$, $c_2(0) = 0$; $\gamma_1 = 6.25 \times 10^6\,\text{s}^{-1}$, $\gamma_2 = 172.4 \times 10^6\,\text{s}^{-1}$; interaction time $T_i = 40\,\text{ns}$, transit time $T = 50\,\text{ns}$. (Adapted from Silverman [8].)

course, in overall intensity, but this need not be a problem for a sufficiently large initial flux of atoms. The pairs $\Delta = \pi, 0$ and $\Delta = \pi/2, -\pi/2$, as first shown by Ramsey, are particularly useful, and lead in the present case to the signals

$$S_1(\pi, 0) = 4e^{-\gamma_1 t}e^{G(T + T_i)} \, \mathrm{Re}\{I_{11}^2 I_{12}^* I_{21}^* e^{i\Omega(T + T_i)}\}, \tag{66b}$$

$$S_1(\pi/2, -\pi/2) = 4e^{-\gamma_1 t}e^{G(T + T_i)} \, \mathrm{Im}\{I_{11}^2 I_{12}^* I_{21}^* e^{i\Omega(T + T_i)}\}, \tag{66c}$$

with comparable expressions for S_2 derivable from (64b).

In Figure 5.11 the variation with frequency of $S_1(\pi, 0)$ and $\rho_{11}(\Delta = 0)$ are superposed in the case of two coupled unstable states (hydrogen $3S$ and $3P$). From the scale of the figure, it is seen that the interference term is approximately one hundred times smaller than ρ_{11} in the vicinity of resonance, and therefore the contrast of the interference pattern is greatly enhanced in the difference signal $S_1(\pi, 0)$. *Ideally*—i.e., in the absence of the "counter-rotating" component of the oscillating field—the interference pattern of $S_i(\pi, 0)$ $(i = 1, 2)$ is symmetric about the resonance frequency. The narrowing of the central portion of the line shape (compared with ρ_{ii}), and therefore the advantage to spectroscopy, are readily apparent. This narrowing becomes more pronounced, as already explained, for longer transit times T between interaction regions, as illustrated in Figure 5.12.

Figure 5.12 also shows the results of choosing the alternative pair of phases, $\Delta = \pi/2, -\pi/2$. The line shape, resembling that of a dispersion curve, is antisymmetric about the resonance frequency $(\Omega = 0)$ at which point $S_i(\pi/2, -\pi/2)$ is ideally null and the slope of the curve is steepest. This, too, is spectroscopically useful, for it can be more advantageous to determine the zero-crossing of a resonance line than to locate the maximum point in a region of near-zero slope. A line shape of the dispersion type is very sensitive to small shifts in the resonance frequency.

5.6. Ion Interferometry and Tests of Gauge Invariance

Upon learning of the diffraction of helium atoms and hydrogen molecules from a crystal, I.I. Rabi allegedly remarked, "They will be diffracting grand pianos next!" While this is unlikely, to say the least, the coherent splitting and subsequent interference of atomic beams, once also a remote possibility, has been achieved and is an actively pursued area of physics research [34]. Since atoms are not charged (like electrons) or penetrate matter (like neutrons), the construction of atom interferometers required first the development of new techniques for simulating necessary optical elements. Recent developments in optics and atomic physics have made possible various types of atom interferometers such as those based on:

(i) wave front splitting at nanometer-sized mechanical structures [35];

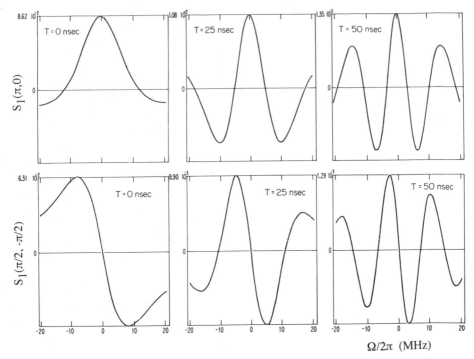

Figure 5.12. Illustration of separated oscillating field interference patterns for different field-free transit times T and for two choices of relative phases $\Delta = (\pi, 0)$ and $(-\pi/2, \pi/2)$. The greater T, the more rapid is the oscillatory structure, and therefore the narrower is the line shape near resonance. The initial amplitudes and decay rates are the same as for Figure 5.11; $V_{12} = 2\pi \times 16$ MHz; $T_1 = 60$ ns. (Adapted from Silverman [8].)

(ii) amplitude division at gratings constructed from standing waves of light [36]; and

(iii) pulsed laser-induced transitions within an atomic "fountain" [37], i.e., a beam of atoms that first rises vertically through an interaction region and subsequently descends slowly through the same region under the action of gravity.

Despite the initial difficulties, an atomic interferometer has a number of potential advantages, not the least of which is that atoms can be produced in beams of high flux. Principally, however, it is on the basis of its high phase sensitivity that atomic interferometers hold great promise as inertial and gravitational sensors in novel experiments to test relativity, search for new physical interactions (such as an intermediate range deviation from Newtonian gravity or so-called "fifth force" [38]), or measure with hitherto unprecedented precision the local acceleration of gravity.

The optical techniques by which neutral atom beam splitters, mirrors, and lenses have been made are generally suitable as well for *charged* atoms, and the development of ion interferometers at some future time can be safely anticipated. Although ions can be controlled by electric fields in the same way that "point" charges like electrons can (and that neutral atoms cannot), it is the internal structure of ions that in part makes them particularly interesting systems to study with the techniques employed in neutral atom interferometry. The interferometry of ions — with internal state labeling — would permit, in a way that has until now been beyond reach, tests of fundamental physical principles related to gauge invariance.

In the first three chapters of this book we discussed various aspects of the Aharonov–Bohm (AB) effect, a purely quantum mechanical phenomenon concerning the interaction of a charged particle with the electromagnetic vector (or scalar) potential under conditions where static electric and magnetic fields are ideally null. Although once the subject of long theoretical and experimental controversy, the AB effect has since been established as a seminal part of quantum theory required by the gauge invariance of the equations of motion. The AB effect illustrates the strange nonlocal influence of topology in physics, as well as the fundamental primacy of potentials (which relate to energy) over fields (which relate to forces) [39]. However, *all* successful AB effect experiments performed to date—whether based on particle beams or mesoscopic metal rings—examined the same type of structureless charged particle, the electron. The reason for this is obvious: coherent electrons are readily available. Nevertheless, novel and conceptually significant modifications of the AB effect can be studied by using composite charged particles with an internal structure, such as ionized atoms [40].

One consequence of the usually assumed "minimal" coupling between particle charge and potential field is that the AB phase shift for single-electron wave packets is independent of the electron energy–momentum–spin state, and depends only on the magnetic flux enclosed by the particle path. Presuming the same coupling holds within a composite structure of bound charges—and *this* is one of the interesting issues worth testing—it can nevertheless follow that the AB phase shift in ions may be observed on *selected excited states*. We will examine this possibility momentarily. Furthermore, under appropriate circumstances, the AB phase shift can be manifested in the resonance fluorescence of the ions, but not—as traditionally detected—in the particle count rate, itself. This optical manifestation of the AB effect is still attributable to the direct coupling of real particles with the vector potential, and is to be distinguished from predicted AB effects of the photon arising from the coupling of virtual electron–positron pairs with an external potential [41].

As an illustration of the new kinds of experiments made possible by ion interferometry, consider the interferometer configuration of Figure 5.13 through which passes a beam of ions of charge q. Each ion is assumed to

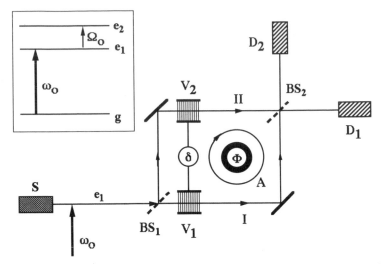

Figure 5.13. The AB experiment employing ion interferometry. A beam of ions, excited into state e_1 by laser, is coherently divided into components passing through one or the other of two coherently oscillating fields V_1 and V_2 (with relative phase δ) before recombining and exiting the interferometer. The effect of confined magnetic flux Φ may appear in either the excited ion count or in the ion resonance fluorescence, depending upon the transitions induced by the oscillating fields.

be a three-level system with ground state $|g\rangle$ and two close-lying excited states $|e_1\rangle$, $|e_2\rangle$ with respective energies ω_1, ω_2 (as usual in units of $\hbar$) and common decay rate Γ. Although the figure shows an interferometer of the Mach–Zehnder type, the beam-splitters and mirrors need not be massive structures; they are merely interaction regions for changing particle momentum or internal state, and can in practice be regions of electromagnetic radiation such as employed in recently developed atomic interferometers with internal state labeling [42]. The ions, excited by laser to lower state $|e_1\rangle$, for example, are split equally at interaction region ("beam-splitter") BS_1 into components that follow paths I or II through the interferometer around a region of confined magnetic flux Φ.

Along the horizontal segment of each path the ions pass through a resonantly oscillating radiofrequency or microwave field that can induce transitions between the two excited states at the Bohr frequency $\Omega_0 = \omega_2 - \omega_1$. The two transition fields V_1, V_2 oscillate coherently with some adjustable, but well-defined, relative phase δ, as analyzed in the preceding section. After traversing the oscillating fields, the ion beam recombines at interaction region ("beam-splitter") BS_2 and passes on to detector D_1 or D_2. We assume the excited state lifetime $1/\Gamma$ to be sufficiently long (as in the case of metastable or Rydberg states) for the ions to traverse the interferometer.

Although, to avoid a linguistic awkwardness, the above description refers to ion "beams" taking paths I and II, one must again bear in mind that, strictly speaking, it is single-ion wave packets that split and propagate through the interferometer. Moreover, in contrast to the standard Ramsey field configuration, where ideally an entire atom or molecule passes sequentially through *both* fields with 100% probability, it is now the case that each component of the ion-wave function passes through only *one* of the fields. Under the conditions of the experiment, the observer does not in general know through which oscillating field a given ion passes. There is thus the potential for interference to occur because of the geometrical (including topological) effects of spatial separation, as well as the temporal (or dynamic) effects of internal state transitions.

After passing interaction region BS_2, the amplitudes of the two components take the general forms

$$\text{PATH I:} \quad (\alpha_{11}e^{i\chi_1^{(1)}}|e_1\rangle + \alpha_{12}e^{i\chi_1^{(2)}}|e_2\rangle)e^{i\gamma_I}, \tag{67a}$$

$$\text{PATH II:} \quad (\alpha_{21}e^{i\chi_{II}^{(1)}}|e_1\rangle + \alpha_{22}e^{i\chi_{II}^{(2)}}|e_2\rangle)e^{i\gamma_{II}}, \tag{67b}$$

where $\chi_J^{(i)}$ is the phase shift of excited state i along path J with possible contributions from excitation, reflection and transmission, and space–time propagation in free space and within the transition regions V_J. The effect of the confined magnetic flux is contained in the nonintegrable phase shift

$$\gamma_J = (q/\hbar c) \int_{\text{Path } J} \mathbf{A} \cdot \mathbf{ds}, \tag{68}$$

where $\mathbf{A}(\mathbf{s})$ is the local vector potential field at some point $\mathbf{s}$ along path J.

The real-valued amplitudes α_{ij} ($i, j = 1, 2$) likewise express the net result of excitation, decay, diffraction, induced transition, and space–time propagation, but they are independent of the vector potential field (or magnetic flux). Their exact expressions can be constructed from the previously derived functions (I_{ij}) of the preceding section, but will not be needed to illustrate the current points of principal interest. Upon recombination, the net amplitude at each detector, for example D_1, takes the form

$$\psi(D_1) \sim (a_{11}e^{i\chi_1^{(1)}}e^{i\gamma_I} + \alpha_{21}e^{i\chi_{II}^{(1)}}e^{i\gamma_{II}})|e_1\rangle + (\alpha_{12}e^{i\chi_1^{(2)}}e^{i\gamma_I} + \alpha_{22}e^{i\chi_{II}^{(2)}}e^{i\gamma_{II}})|e_2\rangle, \tag{69}$$

where, to avoid unnecessary complexity in the mathematical formalism, the additional effects of space–time propagation and decay over the path from BS_2 to D_1 have simply been absorbed in α_{ij} and $\chi_J^{(i)}$.

The arrival of excited ions can be observed in two experimentally distinct ways. One can, in principal, count particles, in which case the detector D_1 is sensitive only to excited states, as, for example, by means of a resonant ionization procedure. On the other hand, one can count the photons in the decay radiation, in which case D_1 represents a photomultiplier and associated electronics. The probability of direct particle detection at D_1 is given by

$$P(D_1) = |\psi(D_1)|^2 \tag{70a}$$

for suitably normalized wave function ψ. By contrast, the fluorescent signal with polarization ε deriving from electric dipole (μ_E) transitions to some final state—here taken for simplicity to be the ground state $|g\rangle$—is obtained from

$$S(D_1) = \text{Tr}\{\rho\mathcal{O}_{\text{det}}(\varepsilon)\}, \tag{70b}$$

where ρ is the density operator of the system

$$\rho = |\psi(D_1)\rangle\langle\psi(D_1)| \tag{71}$$

and $\mathcal{O}_{\text{det}}(\varepsilon)$ is the detection operator

$$\mathcal{O}_{\text{det}}(\varepsilon) = (\boldsymbol{\mu}_E\cdot\boldsymbol{\varepsilon})|g\rangle\langle g|\boldsymbol{\mu}_E\cdot\boldsymbol{\varepsilon})^\dagger, \tag{72}$$

as described in Chapter 4 on quantum beats.

There are a variety of outcomes depending upon the effects of the oscillating fields V_1, V_2, but we shall consider two. Suppose that the transition regions convert incident states $|e_1\rangle$ into pure states $|e_2\rangle$. Then $\alpha_{21} = \alpha_{11} = 0$, and the probability of detecting an excited ion directly is

$$P(D_1) = \alpha_{12}^2 + \alpha_{22}^2 + 2\alpha_{12}\alpha_{22}\cos(\Delta_2 + 2\pi\Phi/\Phi_0), \tag{73a}$$

where

$$\Delta_i = \chi_{II}^i - \chi_I^i \quad (i = 1, 2) \tag{73b}$$

is the phase shift between component waves of state $|e_i\rangle$ due to a difference in optical path length through the interferometer. The ratio of magnetic flux enclosed by the interferometer to the fluxon is

$$\Phi/\Phi_0 = \oint_{\text{Path (II—I)}} \mathbf{A}\cdot\mathbf{ds}/(hc/q) = \gamma_{II} - \gamma_I. \tag{74}$$

The optical signal deriving from the radiative decay of ions is in this case simply proportional to the particle detection probability

$$S(D_1) = |\mu_{2g}|^2 P(D_1), \tag{75a}$$

where

$$\mu_{ig} = \langle e_i|\boldsymbol{\mu}_E\cdot\boldsymbol{\varepsilon}|g\rangle \quad (i = 1, 2) \tag{75b}$$

is the electric dipole matrix element between states $|e_i\rangle$ and $|g\rangle$. It is to be noted that the phase Δ_2 which enters (73a) and (75a) is independent of observation time, since the phase factor $\exp(-i\omega_2 t)$ appears in the amplitudes for both paths. Thus, (73a), (75a) show a stationary manifestation of the AB effect in both the particle and photon counts. (A similar conclusion follows if pure $|e_1\rangle$ states emerge from both transition regions V_J.) A conceptually significant feature of this experiment—in which the electric charge, but *not* the detected bound state, has followed a closed contour about the magnetic flux—is that it tests the state independence of minimal charge coupling and the consistency in formulation of hierarchical quantum equations of motion (a point to be discussed further, below).

Suppose, however, that transition region V_1 produces pure states $|e_1\rangle$ and V_2 produces pure states $|e_2\rangle$. Then $\alpha_{21} = \alpha_{12} = 0$, and it follows from (69) that the probability of particle detection and the optical signal are, respectively,

$$P(D_1) = \alpha_{11}^2 + \alpha_{22}^2, \tag{76a}$$

$$S(D_1) = \alpha_{11}^2|\mu_{1g}|^2 + \alpha_{22}^2|\mu_{2g}|^2 + 2\alpha_{11}\alpha_{22}|\mu_{1g}||\mu_{2g}|\cos(\Delta_{12} + 2\pi\Phi/\Phi_0), \tag{76b}$$

where

$$\Delta_{ij} = \chi_{\text{II}}^j - \chi_{\text{I}}^i \quad (i \neq j). \tag{76c}$$

In this case, the particle count rate shows no quantum interference at all—which is to be expected since the excited state *spatial* paths are known. On the other hand, the AB effect is now imprinted in the optical signal, which is permissible since the exact *temporal* path from ground state to excited state and back to ground state is not known (provided the bandwidth of the detector is wider than the total spectral emission of the excited manifold). The phase Δ_{12} (in which is absorbed any supplementary contribution from the possibly complex-valued electric dipole matrix elements) contains the observation time in the term $(\omega_2 - \omega_1)t = \Omega_0 t$. Thus, the optical signal manifests quantum beats at the Bohr frequency Ω_0 with an initial phase linearly dependent upon the enclosed magnetic flux.

The essential contribution of the two coherently oscillating transition regions is to ensure production of adjustable coherent superpositions of excited states. In an alternative configuration illustrated in Figure 5.14, the separate beam-splitting and transition regions are replaced by two coherently

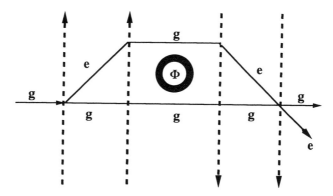

Figure 5.14. Ion AB experiment employing an interferometer consisting of two pairs of coherently oscillating traveling laser fields. Ions are separated according to their internal state by the momentum exchanges attendant to resonant photon absorption and emission.

oscillating counterpropagating sets of laser fields, with two copropagating waves in each set to give, as before, a total of four interaction regions. A beam of two-level ions—one ground level and but one pertinent excited level—enters the interferometer in the ground state. The atomic recoil (analogous to the Kapitza–Dirac effect [43]) produced by resonant absorption and stimulated emission of photons serves to split the ion wave packet coherently in each interaction region into two components depending on the ion state. Emerging from the final laser field are two separated beams, one of ground-state ions and the other of excited-state ions. An excited-state ion, however, could have been generated by a transition (with corresponding deflection) induced by the fourth laser field in the lower ground-state beam (Path I), or could have originated by interaction of a ground-state ion in the upper beam (Path II) with the third laser field. Under the given experimental conditions, one cannot know by which process an emerging ion was put into the excited state. Likewise for the output beam of ground-state ions. The probability that an excited-state or ground-state ion emerges should therefore manifest a quantum interference with a phase dependent on the enclosed magnetic flux—provided that the process of state separation does not destroy the phase memory of the system. Such a system has in fact already been demonstrated in the construction of a neutral-atom Sagnac interferometer [44] in which rotation, rather than magnetic flux, influences the relative phase. (The analogy between rotation and magnetism will be discussed in more detail in the following chapter.)

The above caveat raises an interesting general point regarding separability and coherence. It is well known that a Stern–Gerlach apparatus with a static inhomogeneous magnetic field dephases a superposition of spin substates to the extent that the substates are separated spatially [45]. The argument leading to this conclusion, however, does not apply to dynamic separation methods where time-varying fields do not literally "separate" previously existing states, but rather generate states *in situ* by inducing transitions.

Aside from the practical applications of atom inteferometry—many of which apply as well to ions—there is a particular conceptual interest in examining the AB effect in a charged system with composite structure. In quantum mechanics the structure and interactions of atoms are routinely described—depending upon the question at hand—by different mathematical formulations of the equations of motion, e.g., those of the Dirac, Pauli, and Schrödinger theories. These three formulations are of course intimately related, the second derivable from the first in the approximation of small electron velocity, and the third derivable from the second upon neglect of spin. Nevertheless, each theory must be supplemented by physical assumptions prescribing how observables are to be deduced from the wave function. Furthermore, these theories did not arise in this hierarchical order, but were developed independently.

It has been pointed out years ago [46], in a context different from that under scrutiny here, that there are a number of inconsistencies between the customary formulation of the Schrödinger and Pauli theories and results derived from the Dirac equation. For example, the Schrödinger equation has long been regarded as the nonrelativistic wave equation for a spinless particle, a seemingly reasonable interpretation since one arrives at the Schrödinger equation from a nonrelativistic reduction of the Klein–Gordon, as well as the Dirac, equation. On the contrary, according to Hestenes and Gurtler, the requirement that the Schrödinger theory be identical to the Pauli theory in the absence of a magnetic field necessitates a different interpretation, namely that the Schrödinger equation describe—not a particle without spin—but a particle in a fixed eigenstate of spin. The difference is important and leads to different expressions for the charge current density and a reinterpretation of how the Schrödinger and Pauli currents are associated with momentum and energy. In the words of the authors:

To put it bluntly, everyone to date has been using the wrong expression for charge current density in the Schrödinger theory. Of course there is no way that this error could be revealed directly by experiment, because the only direct experimental means of testing for the existence of a magnetization current is by introducing a magnetic field. But in that case everyone knows enough to discard the Schrödinger theory and use the Pauli or Dirac theories [45a, p. 574].

The interaction of ions with a vector potential field provides another significant circumstance in which the inconsistency between these various formalisms cannot only potentially arise, but be tested. Succinctly stated, it is the replacement—required for the maintenance of gauge invariance by minimal coupling—of momentum $\mathbf{p}$ in the field-free Hamiltonian of a system by the operator $\mathbf{p} - (q/c)\mathbf{A}$ that ultimately leads to the unitary relation (see (1.6a, b))

$$\psi(\mathbf{x}) = \psi_0(\mathbf{x}) \exp\left(i(q/\hbar c) \int^{\mathbf{x}} \mathbf{A} \cdot \mathbf{ds} \right), \tag{77}$$

from which arises the predicted AB phase shift in a space of appropriate topology. Here ψ_0 is a field-free solution of the Dirac equation—or of a *spinless* nonrelativistic wave equation—and ψ the corresponding solution in the presence of a time-independent vector potential. However, the same application of minimal coupling to a nonrelativistic reduction of the Dirac equation with spin-orbit interaction (such as would apply to the internal structure of ions) leads to anomalous gauge-dependent, spin-dependent terms that, if truly present in the Hamiltonian, could conceivably alter the nature of the AB effect in ions through spin-assisted transitions.

There are theoretical reasons for believing these terms to be unwarranted, that correct implementation of gauge invariance must begin at the level of the Dirac equation, and that a proper nonrelativistic reduction of the Dirac equation does not lead in any gauge to nonvanishing spin-dependent terms under the conditions of the AB effect (where $\mathbf{A}$ is ordinarily assumed to be time-independent to avoid classical effects attributable to Faraday induction). Whether or not this reduction leads to gauge-invariant, spin-dependent effects that might show up under other circumstances is not at present a settled issue [47].

Nevertheless, for all its theoretical maturity, physics is ultimately an empirical science. Thus, even apart from the particular issue of hierarchy and gauge invariance discussed above, the uncertainty of an unexpected interaction between a vector potential field and the ionic internal degrees of freedom exists until the question is laid to rest by experiment.

APPENDIX

Oscillatory Field Solution to the Two-State Schrödinger Equation

The oscillatory field solution to (24b) can be expressed in the form

$$\Psi(t) = \begin{pmatrix} \exp(-(\gamma_1/2 + i\omega_1)t) & 0 \\ 0 & \exp(-(\gamma_2/2 + i\omega_2)t) \end{pmatrix} \begin{pmatrix} J_{11} & J_{12} \\ J_{21} & J_{22} \end{pmatrix} \Psi(0), \quad (A1)$$

where

$$J_{11} = \frac{e^{(G-i\Omega)t/2}}{D} [\{1 + \kappa_A^2 e^{2i\omega t}\}\{(1 - \kappa_A K)^2 e^{-\mu t/2} + (\kappa_A + K)^2 e^{\mu t/2}\}$$
$$- 2\{1 - e^{2i\omega t}\}\kappa_A(1 - \kappa_A K)(\kappa_A + K)\sinh(\mu t/2)], \quad (A2a)$$

$$J_{12} = \frac{-2e^{(G-i\Omega)t/2}}{D} [\{1 + \kappa_A^2 e^{2i\omega t}\}(1 - \kappa_A K)(\kappa_A + K)\sinh(\mu t/2)$$
$$- \tfrac{1}{2}\{1 - e^{2i\omega t}\}\kappa_A\{(1 - \kappa_A K)^2 e^{\mu t/2} + (\kappa_A + K)^2 e^{-\mu t/2}\}], \quad (A2b)$$

$$J_{21} = \frac{-2e^{-(G-i\Omega)t/2}}{D} [\{1 + \kappa_A^2 e^{-2i\omega t}\}(1 - \kappa_A K)(\kappa_A + K)\sinh(\mu t/2)$$
$$+ \tfrac{1}{2}\{1 - e^{-2i\omega t}\}\kappa_A\{(1 - \kappa_A K)^2 e^{-\mu t/2} + (\kappa_A + K)^2 e^{\mu t/2}\}], \quad (A2c)$$

$$J_{22} = \frac{e^{-(G-i\Omega)t/2}}{D} [\{1 + \kappa_A^2 e^{-2i\omega t}\}\{(1 - \kappa_A K)^2 e^{\mu t/2} + (\kappa_A + K)^2 e^{-\mu t/2}\}$$
$$+ 2\{1 - e^{-2i\omega t}\}\kappa_A(1 - \kappa_A K)(\kappa_A + K)\sinh(\mu t/2)]. \quad (A2d)$$

The denominator D is

$$D = (1 + \kappa_A^2)(1 + \Delta^2) \quad (A3a)$$

with

$$\Delta^2 = \kappa_A^2 + K^2 + (\kappa_A K)^2, \quad (A3b)$$

κ_A, given by (33a) is the off-diagonal element in the matrix C_A of (32b) which diagonalizes the time-independent part of the Schrödinger equation in the antirotating frame. The function K, defined by

$$K = \frac{iV}{\varepsilon_A^+ + \lambda^- - i\omega},$$ (A4)

is the off-diagonal element in a matrix of the same form as in (32b) which diagonalizes the time-independent part of the Schrödinger equation (34a) transformed back into the rotating frame. $\varepsilon_A^\pm$, given in (34b), are the eigenvalues obtained by diagonalization in the antirotating frame, and

$$\lambda^\pm = -\tfrac{1}{4}(\gamma_1 + \gamma_2) \pm i\mu$$ (A5)

are the eigenvalues obtained by diagonalization in the rotating frame; μ, defined by (35a), is the "precession" frequency.

Upon approximating the radical $(1 + \vartheta^2)^{1/2}$ that appears in μ by 1, one obtains the rotating field solution, i.e., the approximate solution resulting from transformation of the original Schrödinger equation (24b) directly into the rotating frame with subsequent neglect of all antiresonant terms. In particular, it follows that $\lambda^\pm = \varepsilon_R^\pm$, and $K = \kappa_R$, and $\kappa_A = 0$, where the rotating-field eigenvalues are

$$\varepsilon_R^\pm = -\tfrac{1}{4}(\gamma_1 + \gamma_2) \pm iv$$ (A6)

with "precession" frequency v defined by (30a); the associated off-diagonal matrix element is

$$\kappa_R = \frac{\Theta}{\sqrt{1 + \Theta^2} + 1}$$ (A7a)

with

$$\Theta = \frac{2iV_{12}}{G - i\Omega}.$$ (A7b)

(Confer corresponding equations (33a, b).)

References

[1] P.B. Medawar, *The Art of the Soluble* (Methuen, London, 1967), p. 7.

[2] I.I. Rabi, Space Quantization in a Gyrating Magnetic Field, *Phys. Rev.*, **51**, 652 (1937).

[3] J.S. Rigden, *Rabi, Scientist and Citizen* (Basic Books, New York, 1987), pp. 94–95.

[4] W.E. Lamb and R.C. Retherford, Fine Structure of the Hydrogen Atom by a Microwave Method, *Phys. Rev.*, **72**, 2412 (1947); Fine Structure of the Hydrogen Atom, Part I, *ibid.*, **79**, 549 (1950); Fine Structure of the Hydrogen Atom, Part II, *ibid.*, **18**, 222 (1951).

[5] W.E. Lamb, Anomalous Fine Structure of H and He⁺, *Rep. Progr. Phys.*, **14**, 19 (1951).

[6] W.E. Lamb and T.M. Sanders, Fine Structure of Short-Lived States of Hydrogen by a Microwave–Optical Method. I, *Phys. Rev.*, **119**, 1901 (1960).

[7] L.R. Wilcox and W.E. Lamb, Fine Structure of Short-Lived States of Hydrogen by a Microwave-Optical Method. II, *Phys. Rev.*, **119**, 1915 (1960).

[8] M.P. Silverman, Optical Electric Resonance of a Fast Hydrogen Beam, Ph.D. thesis (Harvard University, Cambridge, 1973).

[9] M.P. Silverman and F.M. Pipkin, Optical Electric Resonance Investigation of a Fast Hydrogen Beam. I: Theory of the Atom–RF Field Interaction, *J. Phys. B: Atom. Molec. Phys.*, **7**, 704 (1974); II: Theory of the Optical Detection Process, *ibid.*, **7**, 730 (1974); III: Experimental Procedure and Analysis of $H(n = 4)$ States, *ibid.*, **7**, 747 (1974).

[10] P.M. Stier and C.F. Barnett, Charge Exchange Cross Sections of Hydrogen Ions in Gases, *Phys. Rev.*, **103**, 986 (1956).

[11] C.W. Fabjan, F.M. Pipkin, and M.P. Silverman, Radiofrequency Spectroscopy of Hydrogen Fine Structure in $n = 3, 4, 5$, *Phys. Rev. Lett.*, **26**, 347 (1971).

[12] S.K. Allison, Experimental Results on Charge-Changing Collisions of Hydrogen and Helium Atoms and Ions at Kinetic Energies above 0.2 Kev, *Rev. Mod. Phys.*, **30**, 1137 (1958).

[13] R.F. Stebbings, Charge Transfer, *Adv. in Chem. Phys.*, **X**, 195–246 (1966).

[14] The symbol ε_{ijk}, where i, j, k, can each take values $1, 2, 3$, is ± 1 for, respectively, even and odd permutations of the sequence $1, 2, 3$ and is 0 if any two indices are equal. Thus, for example, $\varepsilon_{123} = \varepsilon_{312} = +1$, $\varepsilon_{213} = \varepsilon_{132} = -1$, and $\varepsilon_{121} = 0$. The symbol δ_{ij} is $+1$ if $i = j$, and 0 if $i \neq j$.

[15] See, for example, M. Born and E. Wolf, *Principles of Optics*, 4th edn. (Pergamon, London, 1970), pp. 30–32; and J.M. Stone, *Radiation and Optics* (McGraw-Hill, New York, 1963), pp. 309–320. For a monochromatic plane light wave represented as a superposition of linearly polarized basis states with (real) amplitudes a_1 and a_2 and relative phase δ, the components h_0, h_1, h_2, h_3 correspond, respectively, to the Stokes parameters defined (up to a constant factor) by the bilinear combinations: $I \sim a_1^2 + a_2^2$, $U \sim 2a_1 a_2 \cos \delta$, $V \sim 2a_1 a_2 \sin \delta$, and $Q \sim a_1^2 - a_2^2$. These parameters are related to the total light intensity (I), the angular orientation and eccentricity of the elliptical motion traced out by the electric vector of the light wave (U and Q), and the handedness or sense of circulation of the electric vector (V).

[16] E.P. Wigner, "Relativistic Invariance and Quantum Phenomena," in *Symmetries and Reflections* (Indiana University Press, Bloomington, 1967), pp. 51–81.

[17] F.E. Hohn, *Elementary Matrix Algebra* 2nd edn. (Macmillan, New York, 1965), pp. 273–306.

[18] M.P. Silverman and F.M. Pipkin, Interaction of a Decaying Atom with a Linearly Polarized Oscillating Field, *J. Phys. B: Atom. Molec. Phys.*, **5**, 1844 (1972).

[19] M.P. Silverman, The Curious Problem of Spinor Rotation, *European J. Phys.*, **1**, 116 (1980) (and references contained therein).

[20] F. Bloch and A. Siegert, Magnetic Resonance for Nonrotating Fields, *Phys. Rev.*, **57**, 522 (1940).

[21] A.F. Stevenson, On the Theory of the Magnetic Resonance Method of Determining Nuclear Moments, *Phys. Rev.*, **58**, 1061 (1940).

[22] W.E. Lamb, Fine Structure of the Hydrogen Atom. III, *Phys. Rev.*, **85**, 259 (1952).

[23] D.A. Andrews and G. Newton, Observation of Bloch–Siegert shifts in the

$2^2S_{1/2}-2^2P_{1/2}$ Microwave Resonance in Atomic Hydrogen, *J. Phys. B: Atom. Molec. Phys.*, **8**, 1415 (1975).

[24] P.A.M. Dirac, *The Principles of Quantum Mechanics*, 4th edn. (Oxford University Pres, London, 1958), p. 148.

[25] A. Messiah, *Quantum Mechanics*, Vol. II (Wiley, New York, 1961), p. 535.

[26] For a discussion of experimental confirmations of the properties of spinor rotation see M.P. Silverman [19], and *And Yet It Moves: Strange Systems and Subtle Questions in Physics* (Cambridge University Press, New York, 1993), Chapter 2.

[27] H. Rauch, A. Zeilinger, G. Badurek, A. Wilfing, W. Bauspiess, and U. Bonse, Verification of Coherent Spinor Rotation of Fermions, *Phys. Lett.*, **54A**, 425 (1975).

[28] S. Werner, R. Colella, A. Overhauser, and C. Eagen, Observation of the Phase Shift of a Neutron Due to Precession in a Magnetic Field, *Phys. Rev. Lett.*, **35**, 1053 (1975).

[29] J. Byrne, Young's Double Beam Interference Experiment with Spinor and Vector Waves, *Nature*, **275**, 188 (1979).

[30] M.P. Silverman, The Distinguishability of 0 and 2π Rotations by Means of Quantum Interference in Atomic Fluorescence, *J. Phys. B: Atom. Molec. Phys.*, **13**, 2367 (1980).

[31] E.D. Bolker, The Spinor Spanner, *Amer. Math. Monthly*, 977 (November, 1973).

[32] See, for example, N.F. Ramsey, *Molecular Beams* (Oxford University Press, Oxford, 1956).

[33] N.F. Ramsey, Experiments with Separated Oscillatory Fields and Hydrogen Masers, *Rev. Mod. Phys.*, **62**, 541–552 (1990). (This paper is the lecture delivered by Ramsey on the occasion of his Nobel Prize.)

[34] See, for example, V.I. Balykin and V.S. Letokhov, Laser Optics of Neutral Atomic Beams, *Physics Today*, **42**, 23–28 (April, 1989); B.G. Levi, Atoms Are the New Wave in Interferometers, *Physics Today*, **44**, 17–20 (July, 1991); M. Sigel and J. Mlynek, Atom Optics, *Physics World*, **6**, 36–42 (February, 1993).

[35] D.W. Keith, C. Ekstrom, Q. Turchette, D.E. Pritchard, An Interferometer for Atoms, *Phys. Rev. Lett.*, **66**, 2693 (1991).

[36] P.L. Gould, G. Ruff, and D.E. Pritchard, Diffraction of Atoms by Light: The Near-Resonant Kapitza–Dirac Effect, *Phys. Rev. Lett.*, **56**, 827 (1986).

[37] M. Kasevich and S. Chu, Atomic Interferometry using Stimulated Raman Transitions, *Phys. Rev. Lett.*, **67**, 181 (1991).

[38] E. Fischbach and C. Talmadge, Six Years of the Fifth Force, *Nature*, **392**, 207 (1992); see also M.P. Silverman, And Yet It Moves: Strange Systems and Subtle Questions in Physics (Cambridge University Press, New York, 1993), Chapter 5.

[39] T.T. Wu and C.N. Yang, Concept of Nonintegrable Phase Factors and Global Formulation of Gauge Fields, *Phys. Rev. D*, **12**, 3845 (1975).

[40] M.P. Silverman, Optical Manifestations of the Aharonov–Bohm Effect by Ion Interferometry, *Phys. Lett. A*, **182**, 323 (1993).

[41] M.P. Silverman, Aharonov–Bohm Effect of the Photon, *Phys. Lett. A*, **156**, 131 (1991).

[42] Ch. J. Bordé, Atomic Interferometer with Internal State Labelling, *Phys. Lett. A*, **140**, 131 (1991).

[43] The Kapitza–Dirac effect is the scattering of a particle beam from a periodic lattice of light created by a standing wave intensity pattern. See, for example, P.L. Kapitza and P.A.M. Dirac, The Reflection of Electrons from Standing Light Waves,

Proc. Cambridge Philos. Soc., **29**, 297 (1933), and the experiments reported in a paper of the same title by H. Schwarz, *Z. Physik*, **204**, 276 (1967).

[44] This interferometer is described in [42]. In a Sagnac interferometer two light (or particle) beams counterpropagate around the same closed path of a rotating interferometer. As a result of the Doppler effect, the recombined beam manifests an optical frequency shift proportional to the rotational angular frequency. It is of interest to note that there is a close analogy between the Sagnac effect and the AB effect with isomorphic connection $q\mathbf{A}/c = m\boldsymbol{\Omega} \times \mathbf{r}$ between angular velocity $\boldsymbol{\Omega}$ and vector potential $\mathbf{A}$. See, for example, M.P. Silverman, Circular Birefringence of an Atom in Uniform Rotation: The Classical Perspective, *Amer. J. Phys.*, **58**, 310 (1990).

[45] G. Baym, *Lectures on Quantum Mechanics* (Benjamin, New York, 1969), p. 327.

[46] (a) R. Gurtler and D. Hestenes, Consistency in the Formulation of the Dirac, Pauli, and Schrödinger Theories, *J. Math. Phys.*, **16**, 573 (1975); (b) D. Hestenes, Spin and Uncertainty in the Interpretation of Quantum Mechanics, Amer. J. Phys., **47**, 399 (1979).

[47] Q. Wang and G.E. Stedman, Spin-Assisted Matter-Field Coupling and Lanthanide Transition Intensities, *J. Phys. B: Atom. Molec. Opt. Phys.*, **26**, 1415 (1993).

CHAPTER 6

The Quantum Physics of Handedness

'Twas brillig, and the slithy toves
Did gyre and gimble in the wabe;
All mimsy were the borogoves,
And the mome raths outgrabe.

Through The Looking-Glass
Lewis Carroll

6.1. Optical Activity of Mirror-Image Molecules

By the end of the second decade of the nineteenth century—long before the discovery of X-rays and the invention of the electron microscope—it would still have been possible, using nothing more than a beam of light and two polarizing crystals (e.g., calcite), to determine that the intrinsic structure of at least some molecules was three dimensional (assuming one believed in molecules then [1]). The tell-tale "optical activity"—a rotation of the plane of polarization of the light upon transmission through the sample as shown in Figure 6.1—does not occur for all molecules, but only for those which cannot be superimposed upon their mirror image. Such a structure, subsequently termed "dissymmetric" by Louis Pasteur but referred to as "chiral" (derived from the Greek for "hand") in current terminology—must necessarily be three dimensional, for a flat object and its mirror image could always be made to superimpose. Reflection, as illustrated in Figure 6.2, reverses the chirality or "handedness" of an object, thereby interchanging, for example, the right and left winding of a helix or screw.

For a molecule to exist in two nonsuperposable mirror-image—or "enantiomeric"—forms it must have no center or planes of symmetry, nor any rotation–reflection axes [2]. A dissymmetric shape is not necessarily devoid of *all* symmetry, however; it may possess pure rotation axes like the C_2 or two-fold rotation axis of a helix. One of the most common manifestations of natural molecular chirality is that associated with tetrahedrally valent carbon atoms bonded to four different substituents (Figure 6.2). Materials composed of distinct enantiomeric forms of the same molecule

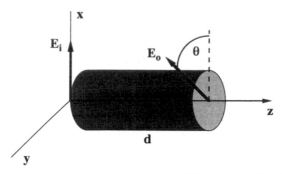

Figure 6.1. Optical rotation of linearly polarized light by a chiral sample. Upon emerging from a sample of length d, the polarization vector $\mathbf{E_o}$ has been turned through an angle θ relative to the incident polarization vector $\mathbf{E_i}$.

have identical bulk chemical and physical properties such as mass density, fusion and vaporization points, and rates of reaction (when chemically combining with *nonchiral* reactants). They also rotate the polarization of a linearly polarized light beam to the same extent per unit length of material, but in *opposite* senses. It is by their optical activity that one can most readily distinguish mirror-image molecules.

In certain cases, however, one could distinguish enantiomeric forms of a substance without any optical apparatus at all, but merely by their aroma. How can the human nose serve as a "chiral detector"? Herein is one

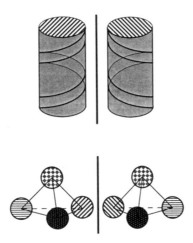

Figure 6.2. Mirror reflection reverses the handedness of chiral objects. A chiral object and its mirror image cannot be superposed. The tetrahedron with four different vertex objects is representative of an "asymmetric" carbon atom, a common chiral building block in chemistry. Some materials, like crystalline quartz, are constructed from *achiral* molecules arranged in a helical structure.

indication of perhaps the most outstanding unsolved problem of the life sciences: the origin of biomolecular homochirality. All life forms—from the lowliest virus to the human body—are built from only one of the two possible enantiomeric forms of its chiral constituents and not (as perhaps one might have expected) from both forms in more-or-less equal measure. An organism may be composed, for example, of right-handed sugar molecules or left-handed amino acids. Indeed, the capacity (if not necessity) to synthesize and consume homochiral molecules is the very hallmark of the living state [3].

Light, too, comes in two basic enantiomeric forms although they are not usually designated as such. These are the left- and right-circular polarizations. From the perspective of classical physical optics—as first explained by Augustin Fresnel in the 1820s—the phenomenon of optical rotation is attributable to circular birefringence, i.e., the difference in indices of refraction (n_l, n_r) for left- and right-circularly polarized light. A circularly polarized wave of angular frequency ω can be regarded as a superposition of two orthogonal linearly polarized waves oscillating with a relative phase of $\pm 90°$ as follows:

$$\mathbf{l} = \frac{1}{\sqrt{2}}(\mathbf{x} + i\mathbf{y}), \tag{1a}$$

$$\mathbf{r} = \frac{1}{\sqrt{2}}(\mathbf{x} - i\mathbf{y}), \tag{1b}$$

(The unit imaginary $\pm i$ can be written as the phase factor $e^{\pm i\pi/2}$.) Here $\mathbf{l}, \mathbf{r}$ are complex forms of the unit vectors representing left- and right-circular polarizations, respectively, for a wave propagating along the z-axis. The unit vectors $\mathbf{x}, \mathbf{y}$ (together with $\mathbf{z}$) are directed along the corresponding axes of an orthogonal *right-handed* coordinate system, i.e., $\mathbf{x} \times \mathbf{y} \cdot \mathbf{z} = 1$. To obtain the physical traveling wave, which is necessarily a real-valued, time-dependent function, one multiplies the expressions in (1a, b) by the phase factor $e^{-i\omega t}$ and then takes the real part. This leads to expressions

$$\mathbf{l} = \frac{1}{\sqrt{2}}(\mathbf{x} \cos(\omega t) + \mathbf{y} \sin(\omega t)), \tag{1c}$$

$$\mathbf{r} = \frac{1}{\sqrt{2}}(\mathbf{x} \cos(\omega t) - \mathbf{y} \sin(\omega t)), \tag{1d}$$

in which the feature of "rotation" is plainly evident. According to standard optical convention, the electric field vector of a left-circularly polarized (LCP) wave rotates toward the left side of an observer facing the light source—with corresponding definition of right-circular polarization (RCP) [4]. We will not need to distinguish by separate symbols the complex and real polarization vectors. The complex form is generally the more convenient to use, since the differential effects of a medium on the two states of polarization can be accounted for simply by appropriate phase factors of the form $e^{i\varphi}$ where, as we will see shortly, φ can be complex-valued if light absorption (or emission) occurs.

Consider, therefore, an x-polarized wave incident upon a sample of transparent optically active material. (See, again, Figure 6.1.) Representing the incident wave by its electric field $\mathbf{E}_i$, and inverting the relations (1a, b), one can write

$$\mathbf{E}_i = \mathbf{x} = \frac{1}{\sqrt{2}}(\mathbf{l} + \mathbf{r}). \tag{2a}$$

Inside the chiral medium, the left- and right-circularly polarized components of the incident wave advance at a phase velocity determined by their respective refractive indices $n_{l, r}$. The recombined wave emerging from the sample after a geometric path length d then has the form

$$\mathbf{E}_0 = \frac{1}{\sqrt{2}}(\mathbf{l}e^{in_l d/c} + \mathbf{r}e^{in_r d/c}) = \frac{1}{\sqrt{2}}(\mathbf{x}\cos\theta - \mathbf{y}\sin\theta) \tag{2b}$$

of a linearly polarized wave rotated by the angle

$$\theta = \frac{(n_l - n_r)\omega d}{2c}. \tag{3}$$

If $n_l > n_r$ the rotation angle is positive and the polarization vector is rotated toward the right side of an observer facing the source. In other words, the rotation is in the sense of the wave with the larger phase velocity.

Optical rotation of linearly polarized light is but one example of optical activity, the assortment of optical responses that can distinguish chiral forms. Circular dichroism is another in which linearly polarized light propagating through a nontransparent chiral substance is converted to elliptically polarized light. (Elliptical polarization is a linear superposition of linear and circular polarizations; the motion of the electric vector projected onto a plane perpendicular to the direction of light propagation traces out an ellipse.) In this case the phenomenon is attributable to the difference in absorption of left- and right-circular polarizations. Both optical rotation and dichroism can be treated together by assigning different complex-valued refractive indices of the form $\tilde{n} = n + i\kappa$ for left- and right-circular polarizations. The absorption coefficient κ then leads to a diminution in amplitude with distance: $e^{i\tilde{n}d/c} = e^{ind/c}e^{-\kappa d/c}$. In addition to the manifestations of optical activity in transmitted light, there are other effects, more difficult to observe, but of conceptual and practical importance, associated with reflected [5, 6] or scattered [7] light. This includes, for example, the difference in reflection of incident LCP and RCP light, and the conversion by reflection of an incident linearly polarized wave into an elliptically polarized wave.

Optics, alone, does not account for the origin of the circular birefringence of chiral media. For this, one must have a microscopic model of the way in which a chiral molecule interacts with electromagnetic radiation. Within the framework of classical physics, a simplified heuristic explanation of optical rotation may be had by examining a system of conducting helices irradiated by linearly polarized electromagnetic waves. Consider first a single helix. The electric field of the incident wave drives charges back and forth along the helical pathway thereby inducing oscillating electric and magnetic dipole moments which radiate secondary (or scattered) waves with their own characteristic dipole patterns. The electric dipole arises because of charge separation along the helix; the magnetic dipole is a consequence of the projected circular motion of the charges in a plane perpendicular to the helical axis. Depending upon the handedness of the helix, the induced moments will be either parallel or antiparallel to one another, and the net electric field of the scattered radiation from both dipoles can therefore take on two different orientations relative to the electric field of the incident wave [8]. The plane of vibration of the resultant transmitted wave, a superposition of the incident and scattered waves, will consequently depend on the sense of the helix. Although, in a system of randomly oriented helical molecules, the extent of optical rotation will depend on the projection of the incident electric field along each helix, the sense of rotation will always be the same for helices of the same handedness.

From the perspective of quantum theory, optical rotation is the outcome of a quantum interference process occurring in the elastic scattering of incident photons by the induced electric and magnetic dipoles of a chiral molecule. There are several interesting features that distinguish this process from other examples of quantum interference we have considered previously. First, limiting this discussion to nonabsorbing molecules, one can remark that the light scattering involves virtual, as opposed to real, processes. A molecule, initially in its ground state, undergoes a nonresonant cyclic transition to an excited state and back again to the ground state by one of several indistinguishable interaction pathways. The apparent violation in energy conservation for each one-way transition is, of course, undetectable, for the final state of the system conserves energy, and any attempt to probe the system during scattering would perturb it and destroy the sought-for interference effect. Second, the various interaction pathways correspond to different time sequences of absorption and emission by the electric and magnetic dipoles. However, since electric and magnetic dipole interactions are connecting the *same* two states (ground and excited), the states of the chiral molecule cannot be eigenstates of well-defined parity, and the process would appear to violate parity conservation. This is a consequence of the *molecular* dissymmetry. (The composite atoms have a center of symmetry.) Yet structural optical activity is, after all, the result of an electromagnetic process—and the laws of electrodynamics, both classical and quantum, are invariant under parity transformations.

The invariance of electromagnetism under both parity (or coordinate inversion $x \rightarrow -x$) and time-reversal ($t \rightarrow -t$) can be demonstrated directly from Maxwell's equations. Under inversion of coordinates the electric and magnetic fields transform as follows: $E(x) \rightarrow -E(-x)$, $B(x) \rightarrow B(-x)$. Likewise, reversing time leads to the transformation: $E(t) \rightarrow E(-t)$,

$\mathbf{B}(t) \to -\mathbf{B}(-t)$. Since all fields of a particular kind, irrespective of the charge or current configuration producing them, must behave in the same way under symmetry transformations, one can understand the above transformations by examining the simple systems of a charged parallel-plate capacitor and a current-carrying solenoid. Under inversion, the plates of the capacitor exchange positions, and the orientation of the electric field is thereby reversed. Since the charges are stationary, they (and the electric field) are unaffected by time reversal. Similarly, coordinate inversion has no effect on either the handedness of the solenoid or the direction of charge flow, in which case the magnetic field is unaffected. Under time reversal, however, charge flows through the solenoid in the opposite direction, and the orientation of the magnetic field is therefore reversed. The net effect of these transformations is to leave Maxwell's equations invariant.

In the following section we will examine more closely optical activity as a quantum interference process and the issue of symmetry. It will be seen that optical activity deriving from chiral molecular structure does not violate the conservation laws of electrodynamics. On the other hand, optical activity in unbound atoms would. Atomic optical rotation and other manifestations of optical activity in atoms have in fact been observed and cannot be accounted for by electromagnetic processes alone, but arise from the weak nuclear interactions [9].

6.2. Quantum Interference and Parity Conservation

The interaction of an electromagnetic wave with a chiral molecule or crystal is in general a complicated process to treat. However, to illustrate simply how just one facet of optical activity, namely optical rotation, arises as a quantum interference process, we adopt a model [10] which discards all but the most essential features of the interaction and reduces the problem to the familiar form of a two-level system. In this unadorned quantum electrodynamic (QED) model, the direction of propagation, wave number, total energy, and occupation number of an incident photon are unchanged; the influence of the medium is manifested exclusively in its effects on the state of the photon's polarization. Since the photon is undeviated in its passage through the sample, the process is known as forward scattering.

Suppose the incident light propagates along the z-axis and is polarized along the x-axis. One can represent this state quantum mechanically by the spinor $\begin{pmatrix} 1 \\ 0 \end{pmatrix}$, where the corresponding spinor $\begin{pmatrix} 0 \\ 1 \end{pmatrix}$ characterizes a y-polarized photon [11]. The complete state of a forward scattered photon (within the framework of our assumptions above) then takes the form $\begin{pmatrix} \phi_1 \\ \phi_2 \end{pmatrix}$ where the complex-valued amplitudes ϕ_i ($i = 1, 2$) are to be determined by solution of the Schrödinger equation with appropriate Hamiltonian. The general form, however, is one which we have encountered in the preceding chapter

$$H\Psi = \begin{pmatrix} \Omega & \varpi \\ \varpi^* & \Omega \end{pmatrix} \begin{pmatrix} \phi_1 \\ \phi_2 \end{pmatrix} = i\frac{\partial \Psi}{\partial t}. \tag{4}$$

All elements of the Hamiltonian matrix are again expressed in angular frequency units. The diagonal elements characterize the energy of the light in the chiral medium which is the same for either state of linear polarization. The off-diagonal elements characterize the interaction (to be specified shortly) which breaks the degeneracy of the polarization states and induces transitions between them. Since no absorption or dissipation occurs in the processes under consideration, the Hamiltonian is Hermitian, and the off-diagonal elements, h_{ij}, in general satisfy the relation $h_{21} = h_{12}^*$ as indicated explicitly in the matrix form of H in (4). The interaction ϖ can be written in the form

$$\varpi = Me^{i\alpha} \tag{5}$$

with real-valued modulus M and phase α.

The chiral molecules and light together comprise a single, closed system in which total energy is conserved and the Hamiltonian is independent of time. Using the relations of Chapter 5 (in particular (5.5) and (5.7a, b)) with the correspondences

$$h_0 = \Omega; \qquad h_1 = \text{Re}(\varpi); \qquad h_2 = \text{Im}(\varpi); \qquad h_3 = 0, \tag{6}$$

one can integrate (4) immediately to obtain the polarization state

$$\begin{pmatrix} \phi_1 \\ \phi_2 \end{pmatrix} = e^{-i\Omega t} \begin{pmatrix} \cos(Mt) & -ie^{i\alpha}\sin(Mt) \\ -ie^{-i\alpha}\sin(Mt) & \cos(Mt) \end{pmatrix} \begin{pmatrix} \phi_1^0 \\ \phi_2^0 \end{pmatrix}. \tag{7}$$

If the incident light is x-polarized, the above state reduces to

$$\begin{pmatrix} \phi_1 \\ \phi_2 \end{pmatrix} = e^{-i\Omega t} \begin{pmatrix} \cos(Mt) \\ -ie^{-i\alpha}\sin(Mt) \end{pmatrix}, \tag{8a}$$

which represents the geometrical polarization vector

$$\mathbf{e} = \cos(Mnd/c)\mathbf{x} - e^{i(\pi/2 - \alpha)}\sin(Mnd/c)\mathbf{y}, \tag{8b}$$

where, as usual, the global phase factor has been suppressed, and the time for a photon with phase velocity c/n to propagate a length d is $t = nd/c$. (There is no distinction to be made in this idealized model between phase velocity and group velocity because the photons are assumed to be in monochromatic plane-wave states.)

In will be demonstrated soon that the matrix element ϖ is a pure imaginary number, with a phase $\alpha = \pm\pi/2$ depending on the light frequency and the sign of a particular product of electric and magnetic dipole matrix elements referred to as the "rotational strength." If the rotational strength is positive and the frequency is below that of the nearest electronic resonance, then $\alpha = \pi/2$, and therefore $\varpi = iM$. The matrix element ϖ changes sign as the light frequency passes through each resonance. Ordinarily, molecular electronic transitions fall in the ultraviolet portion of the spectrum, and one observes optical rotation by means of lower-frequency visible light. Substitution of $\alpha = \pi/2$ in (8b) leads to a polarization vector identical in form to that in relation (2b) with the optical rotation angle now given by

$$\theta = \frac{Mnd}{c}. \tag{9}$$

Although (2b) [or (8b)] and (8a) characterize the same physical state, their interpretations are quite different. In the classical picture of optical rotation the electric vector of the light wave undergoes a continuous rotation as the beam passes through the chiral medium. At any moment the polarization is a well-determined quantity. According to QED, however, the exact polarization of the photon is uncertain until a measurement is performed. A measurement at time t would reveal a state of x-polarization with a probability

$$P_x(t) = \cos^2(Mt) \tag{10a}$$

and a state of y-polarization with corresponding probability

$$P_y(t) = \sin^2(Mt). \tag{10b}$$

Rotating the analyzer (aligned initially along $\mathbf{x}$) through the angle θ of (9) would reveal a polarization state of (8b) with 100% probability.

Comparing (3) and (9) allows one to express the circular birefringence in terms of the QED matrix element

$$n_1 - n_r = \frac{2Mn}{\omega}. \tag{11}$$

Since the two refractive indices for circular polarization must reduce to the mean index n in the absence of a chiral interaction ($\varpi = 0$), and interchange under a change of sign of ϖ, it follows that

$$n_{1,\,r} = n\left(1 \pm \frac{M}{\omega}\right), \tag{12}$$

where the upper sign corresponds to left-circular polarization.

The consistency of the above reasoning, based on a comparison of results from QED and classical optics, can be checked by returning to the Schrödinger equation (4) and diagonalizing the Hamiltonian to obtain the eigenvalues and eigenvectors of the photon in a chiral medium. For the present case of a frequency below resonance where $\alpha = \pi/2$, solution of this simple eigenvalue problem leads to the characteristic frequencies

$$\Omega_{\pm} = \Omega \pm M \tag{13a}$$

and associated vectors (normalized to unit magnitude)

$$\phi_\pm = \frac{1}{\sqrt{2}}\begin{pmatrix} 1 \\ \mp i \end{pmatrix}. \tag{13b}$$

Inspection of relations (1a, b) and (13b) shows that $\phi_\pm$ are states of right- and left-circular polarization, respectively. However, do the eigenvalues in (13a) correlate correctly with the appropriate refractive indices?

Applying the dispersion relation between frequency and wave number

$$\Omega = vk = ck/n, \tag{14}$$

where v is the phase velocity, to the eigenfrequencies in (13a), leads to the expressions

$$\frac{ck}{n_\pm} = \frac{ck}{n} \pm M, \tag{15a}$$

from which the chiral refractive indices can be determined. Since the chiral interaction M is orders of magnitude smaller than an optical frequency (as will be demonstrated shortly), one can solve for $n_\pm$ in the approximation of $M \ll ck$ to find

$$n_\pm = n\left(1 \mp \frac{Mn}{ck}\right) = n\left(1 \mp \frac{M}{\Omega}\right) \tag{15b}$$

in agreement with (12) and (13b) where n_+ is associated with right circular polarization and n_- with left circular polarization. According to our model, there is no energy gained or lost when a photon enters the chiral medium; the diagonal element Ω can then be identified with the vacuum angular frequency ω. The significance of the eigenvectors $\phi_\pm$ is that they propagate through the chiral medium *unchanged* in form and at well-defined phase speeds characterized by their respective indices of refraction.

We examine next the dynamical part of the problem leading to the interaction element $Me^{i\alpha}$. To simplify matters, consider first the interaction of a photon and a single chiral molecule with ground state $|0\rangle$ and spectrum of excited states $|n\rangle$. The Hamiltonian for the total system of radiation and molecule is then the sum

$$H = H_{rad} + H_{mol} + H_{int}, \tag{16a}$$

where the interaction term

$$H_{int} = -\boldsymbol{\mu}_E \cdot \mathbf{E} - \boldsymbol{\mu}_M \cdot \mathbf{B}, \tag{16b}$$

expresses the coupling of the molecular electric dipole moment $\boldsymbol{\mu}_E$ and magnetic dipole moment $\boldsymbol{\mu}_M$ to the electric and magnetic fields of the photon [12]. Explicit expressions for the other terms of the Hamiltonian are not needed; the first leads to the energy of the free radiation field (represented in our problem by the frequency ω) and the second to the molecular energies

Ω_n which we shall assume to be known. As before, the Hamiltonians are in units of $\hbar$, and we will refer to energy and angular frequency interchangeably.

Within the framework of QED, the electric and magnetic fields in (16b) are operators which act on photon states $|\mathbf{k}\lambda\rangle$ of momentum $\hbar\mathbf{k}$ and polarization label $\lambda = 1, 2$ designating the two orthonormal linearly polarized basis vectors $\mathbf{e}^{(\lambda)}$ for each wave vector $\mathbf{k}$. (Strictly speaking, one should write $\mathbf{e}^{(\lambda)}(\mathbf{k})$, but we will not do this to avoid encumbering the notation unnecessarily, since the direction of $\mathbf{k}$—and therefore of $\mathbf{e}^{(\lambda)}$—remains unchanged throughout the scattering.) The total system of radiation and matter can then be represented initially by a state vector of the form $|0; \mathbf{k}1\rangle$ for a molecule in its ground state and one photon of polarization $\mathbf{e}^{(1)}$. From the explicit representation of the radiation fields in terms of creation $a(\mathbf{k}, \lambda)^\dagger$ and annihilation $a(\mathbf{k}, \lambda)$ operators [13]

$$\mathbf{E}(\mathbf{r}) = \sum_{\mathbf{k}\lambda} (2\pi\hbar\omega/V)^{1/2} i\mathbf{e}^{(\lambda)}[a(\mathbf{k}\lambda)e^{i\mathbf{k}\cdot\mathbf{r}} - a(\mathbf{k}\lambda)^\dagger e^{-i\mathbf{k}\cdot\mathbf{r}}], \tag{17a}$$

$$\mathbf{B}(\mathbf{r}) = \sum_{\mathbf{k}\lambda} (2\pi\hbar\omega/V)^{1/2} \frac{i\mathbf{k} \times \mathbf{e}^{(\lambda)}}{k} [a(\mathbf{k}\lambda)e^{i\mathbf{k}\cdot\mathbf{r}} - a(\mathbf{k}\lambda)^\dagger e^{-i\mathbf{k}\cdot\mathbf{r}}], \tag{17b}$$

it is seen that $\mathbf{E}$ and $\mathbf{B}$ create or annihilate one photon at a time. In the above expressions V is the quantization volume—i.e., the volume within which the field modes are defined—which we identify here with the total volume of sample containing chiral molecules. Although the frequency of the incident photon is assumed not to correspond to a resonance of the molecule, the Hamiltonian H_{int} induces transitions to intermediate states of the form $|n; 0\rangle$ in which the incoming photon has been absorbed and $|n; \mathbf{k}1, \mathbf{k}2\rangle$ in which an additional photon has been emitted, the molecule in both cases being in an excited state. These violations of energy conservation are immediately rectified by de-excitation of the molecule and subsequent emission or absorption of a photon to result in the final state $|0; \mathbf{k}2\rangle$. Thus, in the overall energy-conserving process of forward elastic scattering, the molecular state is unchanged and an incoming photon of polarization $\mathbf{e}^{(1)}$ emerges with polarization $\mathbf{e}^{(2)}$.

The probability amplitude for this process is given by perturbation theory to lowest nonvanishing order as

$$V_{1\to 2} = \sum_I \frac{\langle 0; \mathbf{k}2|H_{\text{int}}|I\rangle\langle I|H_{\text{int}}|0; \mathbf{k}1\rangle}{E_0 - E_I}, \tag{18}$$

where the sum is over all possible intermediate states I. The energy (frequency) of the initial state is $E_0 = \Omega_0 + \omega$; the energy of an intermediate state I is either $E_I = \Omega_n$ or $E_I = \Omega_n + 2\omega$ depending on whether the transition was effected by absorption or emission of a photon. Substitution of H_{int} into (18) and evaluation of the resulting expression by means of (17a, b), lead to a sum of four distinct matrix elements which contribute to

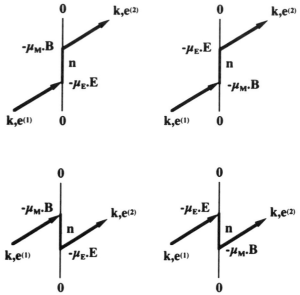

Figure 6.3. Time-ordered diagrams of forward scattering processes contributing to optical rotation. Vertical lines represent the transition of a molecule out of, and back to, its ground state 0 via a virtual transition to intermediate excited state n. Oblique lines represent absorption of a photon with polarization $e^{(1)}$ and emission of a photon with polarization $e^{(2)}$. The diagrams are distinguished by the order of electric and magnetic dipole interactions and by the sequence of absorption and emission.

the change in photon polarization. These are represented diagrammatically in Figure 6.3 in which the time-ordered sequence of events in each diagram proceeds from bottom to top. At each vertex the interaction between molecule and light can be mediated by either an electric or magnetic dipole coupling in which the coupled field either annihilates the incoming photon or creates the outgoing photon. Two interactions (electric or magnetic) of the same kind do not appear in any of the diagrams, for these processes do not change photon polarization.

It may seem particularly strange that a molecule can radiate an outgoing photon from its *ground* state—i.e., *before* the arrival of the incoming photon—but such processes are permitted by QED and must in fact be included if the calculation is to be in accord with experiment. Indeed, the individual vertices of every diagram in the figure violates energy conservation; only *in toto* does a diagram represent a process in accord with physical law. The separate scattering processes of Figure 6.3 are indistinguishable; all that can be observed is that a photon $e^{(1)}$ is incident on a sample of ground-state molecules, and a photon $e^{(2)}$ leaves the same sample. One can, I suppose, disregard the physical interpretation of the diagrams and consider the associated mathematics simply as an exercise in the approximate solution of a differential equation. However, I think this would take away from physics much of the richness of its imagery, for the purpose of physics is not merely to compute, but to understand.

Evaluation of the matrix element (18) leads to the expression

$$V_{1 \to 2} = \left(\frac{4\pi\omega}{V}\right) \sum_n \frac{i\omega \, \mathrm{Im}(\mu_E^{On;\,1} \mu_M^{nO;\,1} + \mu_E^{On;\,2} \mu_M^{nO;\,2}) - \Omega_{n0} \, \mathrm{Re}(\mu_E^{On;\,1} \mu_M^{nO;\,1} - \mu_E^{On;\,2} \mu_M^{nO;\,2})}{\omega^2 - \Omega_{n0}^2}, \quad (19)$$

where the energy interval between ground and excited state is

$$\Omega_{n0} = \Omega_n - \Omega_0 \quad (20)$$

and the dipole matrix elements are defined by

$$\mu_J^{On;\,\lambda} \equiv \langle 0 | \mu_J \cdot \mathbf{e}^{(\lambda)} | n \rangle, \quad (21)$$

with $J = E, M$ and $\lambda = 1, 2$. Equation (19) pertains to a single molecule. In an actual experiment the sample contains a density of η molecules per unit of volume randomly oriented and uniformly distributed over the volume V. One must therefore average (19) over all orientations of the electric and magnetic dipole moments. Then, presuming that each molecule contributes individually to the overall optical rotation (i.e., that there is no cooperative interaction between molecules), one must multiply $V_{1 \to 2}$ by the number of molecules in the sample, ηV. This leads to the final expression for $Me^{i\alpha}$ in the Schrödinger equation (4)

$$\varpi \equiv Me^{i\alpha} = \left(\frac{8\pi\eta\omega^2 e^{i\pi/2}}{3}\right) \sum_n \frac{R_{n0}}{\omega^2 - \Omega_{n0}^2}, \quad (22)$$

where

$$R_{n0} \equiv \mathrm{Im}(\mu_E^{On} \cdot \mu_M^{nO}) \quad (23)$$

is designated the rotational strength for level n.

Although the exact evaluation of R_{n0}, and therefore M, for a real molecule is a difficult calculation, it is worthwhile to make a rough estimate to see the extent of contribution of the chiral interaction to the refractive indices of (12). One can approximate the electric dipole moment by ea_0, the product of electron charge and the Bohr radius of a ground-state hydrogen atom. Similarly, the magnetic dipole moment can be represented by the Bohr magneton, $e\hbar/2mc$, where m is the electron mass. One then obtains $R_{n0} \sim 2.3 \times 10^{-38}$ in cgs units—or $\sim 2.2 \times 10^{-11}$ when divided by $\hbar$—which is more or less an upper limit to actual values, since the induced electric and magnetic dipoles need not in general be parallel nor is the product of the matrix elements necessarily a pure imaginary number. Assuming, further, a sample density of that of water ($\eta \sim 3.3 \times 10^{22}$ molecules/cm^3), a red probe beam ($\lambda = 600$ nm, or $\omega = 3.14 \times 10^{15}$ s^{-1}), and a resonance in the ultra violet ($\lambda = 100$ nm or $\Omega_{n0} = 6\omega$), one obtains $M \sim 1.7 \times 10^{11}$ s^{-1}. The "chiral parameter" in the indices of refraction, (12), is then $M/\omega \sim 5.3 \times 10^{-5}$.

Though small, a chiral parameter of this magnitude is easily measurable and is actually one to two orders of magnitude larger than the corresponding parameters of many naturally occurring chiral molecules. Indeed, one of the recent significant achievements in chiral metrology was the measurement to one part in 10^7 of the difference with which chiral matter reflects LCP and RCP light [14].

We return again to the question of whether or not optical activity—as represented, for example, by the matrix element (22)—violates the conservation of parity, since R_{n0} would be expected to vanish for states of well-defined parity. The fact that the eigenstates of each enantiomeric form of a molecule are not parity eigenstates does not mean, however, that optical activity is necessarily a parity nonconserving process. This question can be answered only by considering the entire system, matter plus radiation. Figure 6.4 illustrates the configuration of a linearly polarized wave incident from the right on a sample of dissymmetric molecules which rotates the polarization clockwise for an observer facing the source—and the mirror image of this process. Upon reflection, the configuration transforms to a linearly polarized wave incident from the left upon the other enantiomeric form of the original molecules; the polarization of the forward scattered light is rotated counterclockwise to an observer facing the source. But this is exactly what one expects to happen. The two enantiomeric forms of a chiral molecule rotate linearly polarized light in opposite senses, the sense of rotation being determined with respect to the direction of light propagation. (There is no other unique direction in the system.) Since both the original and mirror-image processes occur in nature, there is no violation of parity conservation. The mere fact that one can physically separate (or synthesize separately) the two forms of a chiral molecule and carry out optical experiments on only one of these forms does not constitute any violation of physical law.

Suppose, however, that instead of molecules in Figure 6.4, the sample consisted of unbound atoms. Electrodynamically, an atom has a center of

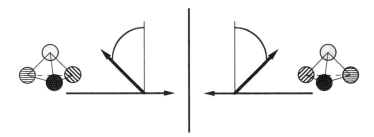

Figure 6.4. Structural optical activity is an electromagnetic process that conserves parity. The mirror inversion (left) of the optical rotation of linearly polarized light transmitted through an entantiomerically pure sample (right) is also an allowable process—namely, a rotation in the opposite sense by a sample of opposite chiral form.

symmetry, since the electrons are bound to the nucleus by the isotropic Coulomb force. One would therefore not expect an atom to come in enantiomeric forms or to rotate the plane of incident linearly polarized light. Nevertheless, atoms have been shown both theoretically and experimentally to be optically active.

One significant outcome of the unification of electromagnetism and the weak nuclear interactions into a single "electroweak" theory, was the prediction of weak neutral currents—in effect, a charge-preserving inter-action between charged particles mediated by the exchange of the Z^0 boson, a neutral particle with mass one hundred times that of the proton mass. Because the range of an interaction is of the order of the Compton wavelength $\lambda_C = \hbar/Mc$ for a mediating particle of mass M, the effects of weak neutral currents in atoms are largely confined to a region $10^{-7}a_0$ about the atomic center. Only electronic S states substantially overlap the atomic nucleus. Thus, as a result of the weak interaction between atomic S electrons with nucleons in the nucleus, electronic S and P states are mixed to a very small extent

$$|S'\rangle \sim |S\rangle + \varepsilon|P\rangle, \tag{24}$$

(with $\varepsilon \sim 10^{-11}$ in hydrogen) and are no longer exact parity eigenstates [15]. (The strength of the coupling grows rapidly with atomic number Z, since the orbital radius of an S electron decreases with Z, and the electron orbital velocity near the nucleus—upon which the weak interaction also depends—increases with Z.)

There is an important distinction, however, between structural optical activity and optical activity attributable to weak neutral currents. Whereas the mirror-image process of the former leads, as illustrated in Figure 6.4, to another process allowed in nature, the mirror-image process of the latter does not occur. The weak nuclear interactions are truly parity violating, and therefore atoms come in but one chiral form; the enantiomeric form does not exist.

Although the chirality of the weak interactions lies essentially in mathe-matical laws rather than in an explicitly visible geometric structure (as in the case of molecules), one can nevertheless construct dynamical quantities that reveal a sense of "handed" motion [16]. Figure 6.5 illustrates the electron probability density current for the hydrogenic $2P_{1/2}$ state

$$\mathbf{J}(\mathbf{r}) = \text{Re}(\langle 2P'_{1/2}|\mathbf{p}|2P'_{1/2}\rangle), \tag{25}$$

where $\mathbf{p}$ is the linear momentum operator, and the designation P' indicates that the actual state has a weak admixture of $2S_{1/2}$ similar to that expressed in (24). S states of other principal quantum numbers are present as well, but their contributions are comparatively negligible. Each streamline, or locus of points everywhere tangent to $\mathbf{J}$, manifests a helical structure as it winds in a preferred sense over a toroidal surface whose axis of rotation (z-axis) is

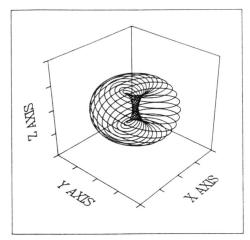

Figure 6.5. Helicity of the hydrogen $2P_{1/2}$ probability density current resulting from weak neutral currents. Shown is the stream line of probability current density **J** with vastly exaggerated mixing coefficient $\varepsilon = 0.5$ and initial point taken to be $(x, y, z) = (6, 0, 0)$ in units of the Bohr radius. (Adapted from Hegstrom et al. [16].)

the quantization axis. The pitch of the helix, determined by the mixing amplitude ε, is greatly exaggerated in the figure for purposes of visibility. In the absence of weak neutral currents, however, ε would be zero, and the corresponding streamlines of a pure $2P_{1/2}$ state of sharp parity would generate circles about the axis of quantization.

6.3. Optical Activity of Rotating Matter

It has been stressed before that optical activity is displayed by chiral materials—structures that cannot be superposed on their mirror image. The required chirality, however, need not always arise from matter alone, but can be a property intrinsic to the larger system encompassing both matter and fields. For example, consider the phenomenon of Faraday rotation in which a sample of *achiral* molecules rotates the plane of linear polarization of a transmitted light beam propagating parallel or antiparallel to a static magnetic field. Although the molecules have no preferential handedness, the magnetic field, which is an axial vector (and not a polar vector like the electric field) imparts a sense of handedness to the system.

Faraday rotation, like natural optical rotation, is parity conserving. Figure 6.6 illustrates the clockwise optical rotation of a light beam propagating to the left, parallel to the magnetic field. Upon reflection, the molecules are unaffected since they are presumed to be achiral, the magnetic field is unchanged (as discussed at the beginning of the previous section), and the

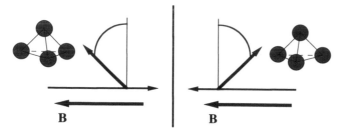

Figure 6.6. The Faraday effect, like structural optical activity, is also an electromagnetic process that conserves parity. The mirror inversion (left) of the optical rotation of linearly polarized light propagating (through an achiral sample) parallel to a static magnetic field is an allowable process—namely, a rotation in the same sense relative to the magnetic field direction.

light, now propagating to the right (antiparallel to the magnetic field), displays a counterclockwise optical rotation. This is exactly what one would expect. Faraday rotation occurs in a fixed sense with respect to the *magnetic field* and not with respect to the direction of light propagation as in the case of structural optical activity. The stark contrast between the Faraday effect and (field-free) optical rotation by intrinsically chiral molecules can be demonstrated by actually reflecting a transmitted light beam back through the sample with a mirror. Upon emerging from a sample of naturally chiral molecules, the net optical rotation will be *zero*. [The rotation along the return path will have reversed the rotation along the forward path because the wave vector of the light is reversed. For both passages, however, the rotation will have occurred in a fixed *sense* (clockwise or counterclockwise) to an observer facing the source.] On the other hand, if the system consists of achiral molecules in a static magnetic field, the optical rotation of the emerging light will be twice that of a one-way passage, since the orientation of the magnetic field has remained the same.

From the perspective of quantum theory, the presence of a static magnetic field **B** splits the degeneracy of magnetic substates of the molecules (Zeeman effect) and leads, by means of perturbation theory, to molecular polarizabilities that depend differently on the magnitude B for left- and right-circularly polarized light. This chiral asymmetry in the polarizabilities translates, through standard relations of electrodynamics, into chirally inequivalent indices of refraction, and hence, by (3) into an optical rotation. We will examine these connections shortly in a different and somewhat unusual context.

There is an insightful analogy known as Larmor's theorem in classical mechanics [17], whereby the motion of a charged particle (charge q, mass m) in a static magnetic field **B** can be analyzed (to a first-order approximation in **B**) as if the particle were in a field-free environment in a rotating reference frame. The hypothetical angular velocity of mechanical roation Ω_L is related to the magnetic field as follows:

$$\Omega_{\mathrm{L}} = \frac{-q\mathbf{B}}{2mc}, \tag{26}$$

where the magnitude of Ω_{L} is the Larmor frequency. The converse of this analogy—namely, that a system undergoing uniform rotation can be treated as if it were subjected to a static magnetic field in a stationary reference frame—has interesting and not widely realized implications for molecular structure and the manifestation of optical activity [18,19]. Indeed, in judging whether or not a system is chiral, one must take account of, as part of the system, not only the material sample and any electromagnetic fields present, but the frame as well.

Let us designate the Hamiltonian and state vector of a quantum system in an inertial reference frame by H and $|\Psi\rangle$ and the corresponding quantities in a rotated reference frame by H' and $|\Psi'\rangle$. Upon rotation of the system through an angle θ about the unit vector $\mathbf{n}$, the two state vectors are related by a unitary transformation

$$|\Psi'\rangle = U|\Psi\rangle = e^{-i\mathbf{J}\cdot\mathbf{n}\theta}|\Psi\rangle, \tag{27}$$

where the generator of rotation, $\mathbf{J}$, is the total angular momentum of the system (and not to be confused with the current density of the previous section). Substitution of (27) into the Schrödinger form of the equation of motion

$$H|\Psi\rangle = -i(d/dt)|\Psi\rangle, \tag{28}$$

yields an equation of identical form in $|\Psi'\rangle$, but with the transformed Hamiltonian

$$H' = UHU^{-1} + iU\frac{dU^{-1}}{dt}. \tag{29a}$$

Use of the explicit expression for U from (27) and assumption of rotational invariance in the inertial frame ($[H, \mathbf{J}] = 0$) results in the Hamiltonian

$$H' = H - \Omega\cdot\mathbf{J}, \tag{29b}$$

where $\Omega = \mathbf{n}(d\theta/dt)$ is the angular velocity of rotation. For a nonrelativistic system in an inertial reference frame, the Hamiltonian can be expressed as the sum of two terms, one for the motion of the center of mass and the other governing the internal dynamics of the system. This same separation can be effected for a rotating system, and it will be hereafter assumed that H' determines the bound-state energy spectrum and that $\mathbf{J}$ refers to the total relative angular momentum of the constituent particles. From (29b) it follows immediately that the energy eigenvectors of the Schrödinger equation are the same in the rotating and inertial frames, but the eigenvalues of the initially degenerate magnetic substates now depend on the magnetic quantum number m_J (i.e., the eigenvalues of J_z where the quantization axis is identified with the axis of rotation). Two substates of the same manifold differing in magnetic quantum number by $\Delta m_J = 1$ are separated by the energy (frequency) interval Ω. In keeping with the aforementioned analogy, the additional term in (29b) has the form of the magnetic Hamiltonian $H_M = (-q/2mc)\mathbf{L}\cdot\mathbf{B}$ for a bound charged particle with orbital angular momentum $\mathbf{L}$—i.e., the isomorphic relation

$$\Omega \Leftrightarrow \frac{q\mathbf{B}}{2mc}, \tag{30}$$

has the opposite sign to that of Larmor's theorem, (26). This is to be expected, since in Larmor's theorem Ω_L was chosen to *cancel* (to first order) the effects of the extant magnetic field. It is also important to note that the quantum mechanical generator of rotations $\mathbf{J}$ is not necessarily identical with $\mathbf{L}$, but can include nonclassical contributions from electron- and nuclear-spin angular momenta.

The fact that rotation formally influences the energy eigenstates of a quantum system as if a magnetic field were present suggests that intrinsically achiral rotating atoms and molecules should display optical activity analogous to the Faraday effect. Suppose such a sample to be irradiated with LCP or RCP light of amplitude E_0 and frequency ω propagating parallel to the axis of rotation. The interaction between the molecule and light is governed principally by the electric dipole Hamiltonian

$$H_E = -\boldsymbol{\mu}\cdot\mathbf{E}_{l,r} = -\tfrac{1}{2}E_0(\mu_\mp e^{i\omega t} + \mu_\pm e^{-i\omega t}), \tag{31a}$$

where

$$\mu_\pm = \mu_x \pm i\mu_y \tag{31b}$$

and the upper and lower signs correspond to LCP and RCP, respectively. In contrast to the analysis of Section 6.2, the rotation-induced optical activity which we are now considering does not arise as an interference between electric and magnetic dipole interactions, and we can neglect here the contribution of the latter which is ordinarily weaker than the former by the ratio v/c where v is the bound electron speed. (When the angular momentum $\mathbf{J}$ derives exclusively from electron and nuclear spin, however—as in the case of ground-state hydrogen hyperfine states—the magnetic dipole coupling plays an important role [20].)

Let us assume that the ground state $|0\rangle$ of the system is an S state. It then follows from first-order time-dependent perturbation theory that the perturbed state vectors $|\Psi_{l,r}\rangle$ contain contributions from virtual transitions to higher P states. Calculation of the electric dipole moments induced by LCP and RCP light

$$\langle\mu\rangle_{l,r} = \text{Re}\{\langle\Psi_{l,r}|\mu|\Psi_{l,r}\rangle\} = \alpha_{l,r}E_{l,r} \tag{32}$$

then leads to the chirally inequivalent polarizabilities

$$\alpha_{l,r} = \frac{1}{\hbar}\sum_n \left[\frac{\Omega_{n0}\mu_{n0}^2}{\Omega_{n0}^2 - (\omega \pm \Omega)^2}\right], \tag{33}$$

where

$$\mu_{n0}^2 \equiv |\langle n| \mu_+ |0\rangle|^2 = |\langle n| \mu_- |0\rangle|^2 \tag{34}$$

and Ω_{n0}, given by (20), is the energy interval between the ground state and excited state n. Only P states with $m_L = 1$ enter expression (33), the contributions from states with $m_L = -1$ having already been included through prior use of the symmetry in relations (34).

Neglecting, for simplicity, the possible distinction between the electric field of the incident wave and the local electric field at a molecular site—an approximation that could, if necessary, be improved by means of the Lorentz–Lorenz formula [21]—one deduces the LCP and RCP dielectric constants and refractive indices from the relation

$$\varepsilon_{1,r} = 1 + 4\pi\eta\alpha_{1,r} = n_{1,r}^2 \tag{35}$$

where η is again the number of molecules per unit of volume. From (35) it then follows that matter in rotation should exhibit a circular birefringence of the form

$$n_1 - n_r \approx \frac{8\pi\eta}{\hbar} \sum_n \left[\frac{\Omega_{n0}\mu_{n0}^2}{(\Omega_{n0}^2 - \omega^2)^2} \right] \omega\Omega, \tag{36}$$

which is linearly proportional to the angular frequency of rotation Ω.

The predicted effect is quite small in comparison to the circular birefringence, (11), of a naturally optically active medium. For the same conditions as before of a sample with the density of water, a resonance at 100 nm, and a red probe beam of 600 nm, (36) leads to a proportionality coefficient of $\sim 2.3 \times 10^{-18}$ s, where we have approximated the dipole matrix element by the product of electron charge and the Bohr radius. The consequences of this circular birefringence, however, are not beyond detection. Thus, for a rotation rate of 100 Hz (which is probably close to the upper limit of what is achievable in the laboratory) and a total path length of 10 m (obtained by multiple reflection of the light through the sample), one could expect an optical rotation of about 4×10^{-6} degrees. (Note that multiple passage of the light is helpful here, since the phenomenon is analogous to the Faraday effect.) Use of a nonresonant ultraviolet probe beam could enhance the signal by one to two orders of magnitude.

Established techniques such as photoelastic modulation and synchronous detection can detect Faraday rotations with a sensitivity of 10^{-4} degrees, while recently developed laser polarimeters, designed for the study of optical activity associated with parity violations in atoms, can detect optical rotations at the level of 10^{-6} degrees with an expectation of improvement by several orders of magnitude [22].

The phenomenon of rotational optical activity, analyzed above within the framework of quantum mechanics, is amenable, as well, to a classical mechanical interpretation [23] which affords a visual image of the effect of classical forces on the orbits of the bound electrons. According to this viewpoint, it is principally the Coriolis "pseudo"-force that acts differently on particles circulating in clockwise or counterclockwise orbits as driven by the electric field of incident circularly polarized light. This leads to chirally inequivalent orbital radii, and therefore to different polarizabilities and refractive indices.

It is instructive to examine this point of view more closely in the simple case of a model atomic system with single particle of mass m and charge q in an isotropic harmonic oscillator potential $U(\mathbf{r}) = \frac{1}{2}m\Omega_0^2 r^2$ with angular frequency of oscillation Ω_0. If the particle is subjected to an electromagnetic plane wave $\mathbf{E}(t)$ of angular frequency ω propagating parallel to the rotation axis of a frame rotating with angular velocity $\boldsymbol{\Omega}$, it experiences an effective force [24].

$$\mathbf{F}_{eff} = \mathbf{F} - 2m(\boldsymbol{\Omega} \times \mathbf{v}) - m\boldsymbol{\Omega} \times (\boldsymbol{\Omega} \times \mathbf{r}), \qquad (37a)$$

where the "true" force, determined in an inertial frame, is

$$\mathbf{F} = -\nabla U(\mathbf{r}) - q\mathbf{E}\cdot\mathbf{r}. \qquad (37b)$$

The second and third terms in (37a) will be recognized as the Coriolis and centrifugal "pseudo"-forces.

Solution of Newton's equation of motion

$$m\frac{d^2\mathbf{r}}{dt^2} = \mathbf{F}_{eff} \qquad (38)$$

for the magnitudes of the steady-state coordinates $\mathbf{r}_l$, $\mathbf{r}_r$ produced by LCP and RCP waves, respectively, leads to the polarizabilities

$$\alpha_{l,r} = \frac{(q^2/m)}{\Omega_0^2 - (\omega \pm \Omega)^2} \qquad (39)$$

and ultimately to the circular birefringence

$$n_l - n_r = \left(\frac{8\pi\eta q^2}{m}\right)\frac{\omega\Omega}{(\Omega_0^2 - \omega^2)^2}. \qquad (40)$$

In both the quantum and classical analyses the plane of polarization is rotated clockwise for an observer facing the light source.

One can formally correlate the classical expressions in (39) and (40) with the quantum mechanical expression of (33) and (36) by identifying the classical dipole moment qr with the matrix element μ_{no} and the oscillation frequency Ω_0 with the resonance frequency Ω_{no} (for a particular state n in the summation), and equating twice the potential energy $m\Omega_0^2 r^2$ to $\hbar\Omega_{no}$.

From (30) it is seen that the magnetic analogue of the Coriolis force is the Lorentz force

$$\mathbf{F}_M = (q/c)\mathbf{v} \times \mathbf{B} \qquad (41)$$

and one could account classically for the Faraday rotation by a formally identical mathematical analysis. There is an important distinction between

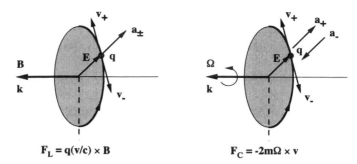

$$F_L = q(v/c) \times B \qquad\qquad F_C = -2m\Omega \times v$$

Figure 6.7. Effects of Lorentz and Coriolis forces on a charged particle driven by an incident traveling wave **E** of left circular polarization. Irrespective of the sign of charge q, the Lorentz force of a magnetic field **B** parallel to the wave vector of the light is directed radially outward. The Coriolis force, attributable to a frame rotating with angular velocity Ω parallel to the wave vector of the light, is directed outward for a positive charge and inward for a negative charge. The instantaneous particle velocities and accelerations are $v_\pm$ and $a_\pm$ where "$\pm$" specifies the sign of charge.

the two forces, Lorentz and Coriolis, however, which illustrates the limitations of the analogy between magnetism and rotation. Note that the chiral polarizabilities (39) are independent of the sign of charge—for both the Faraday effect and rotational optical activity—although the Coriolis and Lorentz forces *differ* in this respect. Consider first the magnetic case illustrated in the left-hand side of Figure 6.7. If the charge q is positive, it is driven by an incident LCP wave (traveling parallel to the magnetic field) to orbit counterclockwise as viewed by an observer facing the source. The Lorentz force then accelerates the particle radially *outward*. If the charge is negative, the LCP wave drives it clockwise, but the Lorentz force, whose direction is also reversed, again accelerates the particle radially *outward*. In either case the LCP wave leads to an orbital radius, and therefore polarizability and refractive index, larger than those in the absence of the magnetic field. Correspondingly, a RCP wave leads to a smaller refractive index, with the result that the birefringence (40) is positive.

Since the Coriolis force, unlike the Lorentz force, is not proportional to electric charge, it accelerates countercirculating positive and negative particles in opposite radial directions as shown in the right-hand side of Figure 6.7. From relation (30), it is seen that reversing the sign of the charge in a magnetic configuration actually corresponds to reversing the angular velocity of the rotating reference frame in the present circumstance. We are considering, however, a fixed sense of rotation. One might intuitively think, therefore, that by the foregoing argument the Coriolis force should lead to $n_l > n_r$ for a positive charge and $n_r > n_l$ for a negative charge—but this is not so. With account taken of all contributions to (37a), it is found that the sign of q determines the relative orientation (parallel or antiparallel) of $E_{l,r}$ and the corresponding $r_{l,r}$, but not the magnitude of $r_{l,r}$.

Looked at from an inertial frame, the inequivalent action of LCP and RCP radiation is effectively attributable to the Doppler effect. A LCP wave of frequency ω propagating parallel to the axis of a frame rotating with angular frequency Ω is perceived to have the frequency $\omega + \Omega$ by an inertial observer. Similarly, the inertially measured frequency of the corresponding RCP wave would be $\omega - \Omega$. Thus, LCP and RCP waves of the same frequency in the rotating reference frame have different frequencies in an inertial reference frame in which case the chiral asymmetry expressed by (39) can be interpreted as a consequence of the frequency dispersion of the polarizability function. For a fixed sense of mechanical rotation, only the sense of rotation of the electric field of the incident light matters, and not the sign of charge.

In general, theoretical analyses of terrestrial atomic and molecular systems ordinarily take for granted at the outset that the frame of reference is inertial, and actual experiments are usually executed under such conditions that this assumption is thought to be adequate. In point of fact, of course, the Earth is not an inertial frame, and the manifestations of the Earth's rotation on *macroscale* systems, both mechanical (e.g., Foucault pendulum) and electromagnetic (e.g., Sagnac effect [25]), have been known for a long time. Only relatively recently, however, has it been possible by means of neutron interferometry to demonstrate the effect of the Earth's gravity and diurnal rotation on a quantum system [26]. Such experiments, which employ beams of free neutrons, raise interesting questions regarding the influence of the Earth's spin on the optical properties of bound-state systems. Would it be possible, for example, to observe optical activity induced in materials by the Earth's rotation where $\Omega/2\pi = 1.2 \times 10^{-5}$ Hz? The corresponding optical rotations would be smaller than those calculated for a rapidly rotating laboratory turntable by some seven orders of magnitude, and are currently beyond the sensitivity of any technique known to the author. Nevertheless, the field of polarimetry is advancing, and developments in ring-laser interferometry, in particular, give cause for a cautiously optimistic response [27].

A ring-laser interferometer is a self-excited optical oscillator in which two counterpropagating beams can be made to interfere upon exiting through a mirror. If the ring is subject to a nonreciprocal effect on the two beams, the interference fringes will be shifted. One such cause would be a circularly birefringent sample—either natural or rotationally induced—in one arm of the interferometer. Recent analysis of ring laser performance [28] showed that a ring of 1 m², operating at a single mode at 633 nm (the red line of a helium–neon laser) with state-of-the-art dielectric mirror coatings, ought to be able to detect nonreciprocal contributions to the birefringence (of a sample of length 10 cm) of $\Delta n \sim 10^{-18}$. This is of the same order of magnitude as the theoretical anisotropy realizable by the Earth's rotation for nonresonant radiation with "detuning" parameter $\omega/\Omega^0 \sim 0.95$. Ring-laser interferometers of this projected sensitivity are under construction at various laboratories in North America, Europe, and New Zealand.

It is, however, often the case that the gulf between pure theory and the outer limits of experimental implementation is not to be bridged easily. Experimental difficulties are always likely to arise—some to be surmounted only by unforeseen advances in technology. Of *this*, though, one can be sure: a new technique which substantially improves experimental resolution will assuredly discover new and worthwhile things.

References

[1] Skepticism over the atomic theory of matter by reputable scientists existed into the twentieth century, i.e., until the conclusive experiments in 1908–1909 on Brownian motion by Jean Perrin based on the theoretical predictions of Einstein (1905–1906). See, for example, G.L. Trigg, *Crucial Experiments in Modern Physics* (Crane, Russak, New York, 1975), Chapter 4.

[2] Optically active substances can also be constructed from mirror-inequivalent arrangements of achiral molecules. Crystalline quartz is one such example; repeating units of silicon dioxide wind in helical fashion (with left or right circulations) about the optic axis. Unlike substances composed of intrinsically chiral molecules, however, the chirality, and therefore the optical activity, vanish when these "enantiomorphic" forms are melted or dissolved in solution. Thus, fused quartz exhibits no optical activity.

[3] Thorough discussions of the problem of biomolecular homochirality may be found in *Origins of Optical Activity in Nature*, edited by D.C. Walker (Elsevier, Amsterdam, 1979), and in the special issue of *Chiral Symmetry Breaking in Physics, Chemistry, and Biology of Biosystems*, **20**, No. 1 (1987).

[4] It is a common error to think that the electric vector of circularly polarized light traces out a circle in time. Since the light wave is advancing as the electric vector is rotating, the actual locus of points traced out would resemble something like a twisted ribbon.

[5] M.P. Silverman, Reflection and Refraction at the Surface of a Chiral Medium: Comparison of Gyrotropic Constitutive Relations Invariant or Noninvariant under a Duality Transformation, *J. Opt. Soc. Amer. A*, **3**, 830 (1986). The controversy on the phenomenological description of optical activity resolved theoretically in this paper is described in detail in M.P. Silverman, *And Yet It Moves* (Cambridge University Press, New York, 1993), Chapter 4.

[6] M.P. Silverman, N. Ritchie, G.M. Cushman, and B. Fisher, Experimental Configurations Employing Optical Phase Modulation to Measure Chiral Asymmetries in Light Specularly Reflected from a Naturally Gyrotropic Medium, *J. Opt. Soc. Amer. A*, **5**, 1852 (1988).

[7] L.D. Barron, *Molecular Light Scattering and Optical Activity* (Cambridge University Press, New York, 1982).

[8] For a given observation direction, the electric fields of waves radiated by electric and magnetic dipoles are mutually perpendicular. However, depending on whether the two fields oscillate in phase or 180° out of phase, the superposition yielding the net scattered wave may have one or the other of two orientations differing

by $90°$. See, for example, E. Hecht and A. Zajac, *Optics* (Addison-Wesley, Reading, MA, 1974), pp. 258–260.

[9] M-A. Bouchiat and L. Pottier, Optical Experiments and Weak Interactions, *Science*, **234**, 1203 (1986).

[10] E.A. Power and T. Thirunamachandran, Optical Activity as a Two-State Process, *J. Chem. Phys.*, **33**, 5322 (1971).

[11] The quantum optical attribute corresponding to circular polarization is technically designated "helicity," the projection $\mathbf{S}\cdot\mathbf{n}$ of the photon angular momentum $\mathbf{S}$ (in units of $\hbar$) onto the direction of its linear momentum $\mathbf{k} = \mathbf{n}k$. Since the photon is a massless spin-1 boson, this projection may have the two values ± 1, which corresponds, respectively, to LCP and RCP states.

[12] One might wonder how the interaction Hamiltonian of (16b) relates to the standard Hamiltonian $H = (\mathbf{p} - q\mathbf{A}/c)^2/2m + q\varphi$ obtained by "minimal coupling" of a charged particle (with linear momentum $\mathbf{p}$) to an electromagnetic vector and scalar potentials $(\mathbf{A}, \varphi)$ in an explicitly gauge-invariant way. This question is by no means a trivial one, and has led to repeated discussion in the physics literature even though the problem was effectively resolved long ago. See, for example, E.A. Power and S. Zienau, Coulomb Gauge in Non-Relativistic Quantum Electrodynamics and the Shape of Spectral Lines, *Trans. Roy. Soc. (London) A*, **251**, 427 (1959). In brief, the electric and magnetic dipole interaction terms correspond to the terms $-q\mathbf{A}\cdot\mathbf{p}/mc + e^2A^2/2mc^2$ in the expansion of the above Hamiltonian after implementation of an appropriate gauge transformation.

[13] The creation and annihilation operators for each mode $(\mathbf{k}\lambda)$ of the radiation field satisfy the standard boson commutation relations $[a(\mathbf{k}, \lambda), a(\mathbf{k}', \lambda')^\dagger] = \delta(\mathbf{k} - \mathbf{k}')\delta_{\lambda\lambda'}$ and act upon a photon state containing N photons in a given mode as follows: $a|N\rangle = N^{1/2}|N - 1\rangle$, $a^\dagger|N\rangle = (N + 1)^{1/2}|N + 1\rangle$. In QED one ordinarily starts with a Fourier representation of the vector potential $\mathbf{A}$ in terms of the creation and annihilation operators and derives the forms of the transverse radiation fields $\mathbf{E}$ and $\mathbf{B}$ from the relations $\mathbf{E} = -\partial\mathbf{A}/c\,\partial t$ and $\mathbf{B} = \nabla \times \mathbf{A}$. A good introduction to the theory of the quantized radiation field is given by G. Baym, *Lectures on Quantum Mechanics* (W.A. Benjamin, New York, 1969), Chap. 13; and by E.A. Power, *Introductory Quantum Electrodynamics* (American Elsevier, New York, 1964).

[14] M.P. Silverman, J. Badoz, and B. Briat, Chiral Reflection from a Naturally Optically Active Medium, *Optics Lett.*, **17**, 886 (1992).

[15] M.A. Bouchiat and C.C. Bouchiat, Weak Neutral Currents in Atomic Physics, *Phys. Lett.*, **48B**, 111 (1974).

[16] R.A. Hegstrom, J.P. Chamberlain, K. Seto, and R.G. Watson, Mapping the Weak Chirality in Atoms, *Amer. J. Phys.*, **56**, 1086 (1988).

[17] H. Goldstein, *Classical Mechanics*, 2nd edn. (Addison-Wesley, Reading, MA, 1980), pp. 232–235.

[18] M.P. Silverman, Rotational Degeneracy Breaking of Atomic Substates: A Composite Quantum System in a Noninertial Reference Frame, *Gen. Relativity and Gravitation*, **21**, 517 (1989).

[19] M.P. Silverman, Rotationally Induced Optical Activity in Atoms, *Europhys. Lett.*, **9**, 95 (1989).

[20] M.P. Silverman, Optical Activity Induced by Rotation of Atomic Spin, *Nuovo Cimento*, **14D**, 857 (1992).

[21] M. Born and E. Wolf, *Principles of Optics*, 4th edn. (Pergamon, Oxford, 1970), p. 87.

[22] S.C. Read, M. Lai, T. Cave, S.W. Morris, D. Shelton, A. Guest, and A.D. May, Intracavity Polarimeter for Measuring Small Optical Anisotropies, *J. Opt. Soc. Amer. B*, **5**, 1832 (1988).

[23] M.P. Silverman, Circular Birefringence of an Atom in Uniform Rotation: The Classical Perspective, *Amer. J. Phys.*, **58**, 310 (1990).

[24] We have again separated the center of mass motion and the internal or relative motion. Equation (37) specifies the force on the "relative" particle in a two-particle system—i.e., the hypothetical particle whose mass is the reduced mass $m_1 m_2/(m_1 + m_2)$ and whose velocity is the relative velocity $v_2 - v_1$. If the mass of one particle is much greater than that of the other, then the motion of the bound particle with smaller mass is virtually the same as that of the "relative" particle.

[25] The Sagnac effect is discussed in [44] of Chapter 5.

[26] J.L. Staudenmann, S.A. Werner, R. Colella, and A.W. Overhauser, Gravity and Inertia in Quantum Mechanics, *Phys. Rev. A*, **21**, 1419 (1980).

[27] M.P. Silverman, Effect of the Earth's Rotation on the Optical Properties of Atoms, *Phys. Lett. A*, **146**, 175 (1990).

[28] G.E. Stedman and H. R. Bilger, Could a Ring Laser Reveal the QED Anomaly Via Vacuum Chirality?, *Phys. Lett. A*, **122**, 289 (1987).

Index

AB–EPR effect   68, 73, 75
AB–HBT effect
  amplitude splitting
    interferometer   79–80
  wavefront splitting
    interferometer   75–76
Aharonov–Bohm (AB) effect   10–13,
    178–179
  Bayh experiment   15–17
  bound-state   23–24
  Hitachi experiment   17–19
  in mesoscopic ring   54–55
  with ions   172–176
Aleksandrov, E.B.   100
angular momentum
  canonical   44
  kinetic   43–44
antibunching   66, 80
anticommutator   107
anyon   49
autocorrelation function   107

Back–Goudsmit effect   131
Bell inequalities   122
biomolecular homochirality
    186
Bloch–Siegert shift   152
Bohr correspondence principle   14
Bohr frequency   102
Bohr magneton   195
Bohr radius   26, 31, 34
Bohr, Niels   21
Bose–Einstein statistics   50, 93
boson   49
brightness   91–93
broadband approximation
    105–109

broken symmetry   23
bunching   66

canonical momentum   10–11
Casimir operator   86
centrifugal force   203
chaotic source   67
charge exchange interaction
    136–137
chiral parameter   195
chirality   36, 184
circular birefringence   187,
    202–203
circular dichroism   187
circular polarization   186
coherence
  area   84, 92
  in density matrix   112, 114
  length   5–6, 54
  time   5
commutation relations
  angular momentum   44
  coordinate momentum   60
Compton wavelength   4
Cooper pair   18
Coriolis force   203–204
correlated two-port states   86
correlation function   63
correlation time   108
COW experiment   85
cross-correlation   83–84, 88–89
cyclic transition   159
cyclotron frequency   38

de Broglie wavelength   4
decay operator   108, 146

degeneracy
  accidental 33
  Stark 25
  Zeeman 39
degeneracy parameter 84, 92
density operator (or density
    matrix) 83, 88, 106, 110, 143,
    146
detection operator 109
diamagnetic interaction 36
Dirac, P.A.M. 10, 96
dispersion relation 192
divergence angle 5
Doppler broadening 103, 126, 130
dynamical observable 41

effective Hamiltonian
  light shift 116
  Stark shift 117
eigenvalue problem 141
Einstein, A. 59, 67
electric dipole interaction 24, 107,
    145, 192
electroweak theory 197
elliptical polarization 187
enantiomeric forms 184
energy operator 43
entanglement 61, 121
EPR paradox 59–60, 121

Faraday effect 198–199
Faraday's law of induction 46
Fermi–Dirac statistics 49, 85, 93
fermion 49
Feynman, Richard vii, x, 1–3, 14
fine structure 145
fine structure constant 145
fluctuation (of light intensity) 63
fluxon 13
forward scattering 194
Fresnel, Augustin 186

Gabor, D. 93
gauge function 42
gauge transformation 9–42
Gaussian pulse profile 113
gravity (effect on quantum
    interference) 85
group
  abelian 32

homotopy 52
rotation-reflection 36, 39
S_2 36
SO(2) 32
SO(3) 33
SO(4) 33
SU(1, 1) 23, 27
SU(n) 86
V 32, 36, 39
group theory 39–40

Hamiltonian (as time–evolution
    operator) 43
Hanle effect 114–116
harmonic oscillator 31, 35, 38, 41
HBT correlation 61
hierarchy of quantum
    theories 177–178

intensity correlation 63
intensity interferometer 61–62, 65
interaction representation 106, 144
ion interferometry 170–173

Kapitza–Dirac effect 177
kinetic momentum 10–11

Lamb–Retherford
    experiment 130–131
Landau levels 38
Landé factor 119
Larmor frequency 35, 200
Larmor's theorem 199, 201
level shift 107, 116
linear-absorption
    approximation 104
Lorentz force 9, 203

Mach–Zehnder
    interferometer 79–80,
    173
magnetic dipole interaction 35,
    117, 192
magnetic flux 10
  quantization 17–19
magneto-resistance 55
Mathieu equation 27, 37
Mathieu function 27–28, 30
matrix diagonalization 141

Maxwell's equations   8
Maxwell, James Clerk   9–10
mesoscopic ring   53
mixture (of states)   102
modulation depth   111
Möllenstedt biprism   1, 17
motional electric field   131
multiple-valued wave function   47

nanotip   94
neoclassical radiation theory   105
nonintegrable phase factor   15
nonlocality   14

optical activity   184
optical electric resonance (OER)   132
optical pumping   102
optical rotation   184–185
oscillating-field theory   142, 147

paramagnetic interaction   35
parity operator   23, 25
parity violation (in atoms)   189,
    197–198
Paschen–Back effect   131
Pauli exclusion principle   21
Pauli matrices   139–140
pertubation theory   26–104
phase sensitivity (of
    interferometer)   89–90
Poisson statistics   95
polarizability   25
precession
    of atomic coherence   115
    of system vector   143–144, 151
preparation factor   111, 112, 113
projection operator   106
pumping time   107
Purcell, E.   64

quantum beat
    and energy indeterminacy   101
    and resonance   159–163
    in entangled states   122–123
    laser-induced   104
    multiatom   126
    regeneration   114–115
    Zeeman   119–121
quantum boost   102

quantum interference
    distinct types   68
    electron build-up   3
    time domain   102
    virtual processes   194
quantum "mysteries"   8
quasi-stationary state   147

Rabi "flopping" formula   130, 147
Rabi, I.I.   130
raising/lowering operators   87
random-walk   114
reflection amplitude   70
reflection operator   25–26
resonance
    lineshape   130, 135–137
    narrowing   69–170
    retarded time   125
ring-laser interferometer   205–206
rotating-field theory   142, 149
rotation operator   143, 200
rotational strength   195
Rydberg state   104, 117

Sagnac
    effect   183, 205
    interferometer   177
saturation parameter   112, 115
scalar potential   9
second-quantization   81
separated oscillating fields   164–166,
    169
shot noise   67
signal-to-noise ratio   95–96
Silverman–Pipkin shift   151
spectral density   107
spin angular momentum   7
spin-orbit coupling   117
spinor rotation   144, 156
spinor spanner   164
squeezed light   67
Stark effect   24, 117, 131, 132
stimulated emission   105, 108
Stokes's parameters   140, 181
Stokes's theorem   9, 42, 77, 175
SU(2) algebra   86–87

three-level atom   109
time-ordered diagram   123, 194

time-reversal
   of electromagnetic fields 188
   of paths 56
transmission amplitude 70
two-level system 129, 139, 190
two-port states 85–86
two-slit interference 1–8

unitarity condition 70

variance 64, 67, 84, 95
vector potential 9

virtual process 188
visibility function 6

wave noise 67
wave packet 4
weak-pumping approximation 104
winding number 51–52, 55

Z^0 boson 197
Zeeman effect 35, 117, 131, 199
zero-point energy 22